GUY P. HARRISON

RACE AND REALITY

WHAT EVERYONE SHOULD KNOW
ABOUT OUR BIOLOGICAL DIVERSITY

Prometheus Books

59 John Glenn Drive
Amherst, New York 14228–2119

Published 2010 by Prometheus Books

Inquiries should be addressed to
Prometheus Books
59 John Glenn Drive
Amherst, New York 14228–2119
VOICE: 716–691–0133
FAX: 716–691–0137
WWW.PROMETHEUSBOOKS.COM

14 13 12 11 10   5 4 3 2 1

Library of Congress Cataloging-in-Publication Data

Harrison, Guy P.
    Race and reality : what everyone should know about our biological diversity / by
Guy P. Harrison.
        p. cm.
    Includes bibliographical references and index.
    ISBN 978–1–59102–767–6 (pbk. : alk. paper)
    1. Race. 2. Physical anthropology. 3. Human population genetics. 4. Biodiversity.
I. Title.

GN269.H37 2009
305.8—dc22

                                                                                    2009032860

Printed in the United States of America on acid-free paper

For Sheree

CONTENTS

FOREWORD

"Race," I tell my students, "is the intersection of difference and meaning." In other words, race cannot be discovered by examining human differences, since all people are different. Race, instead, is constructed by deciding that certain differences are more important than others—that somehow the difference between a Norwegian and a Saudi Arabian is not as significant as that between the Saudi Arabian and a Sudanese; or that the difference between the Norwegian and a German is not as significant as that between the German and a Slovenian.

Race is an act specifically of naming difference, a cultural decision that some differences are worth acknowledging and others are worth ignoring. Consequently, we don't discover race "out there"; rather, we construct it "in here" (you can't see it, but I'm pointing to my head).

Other professors have their own ways of talking about it. Some begin with Donna Haraway's observation that "race is a verb." In other words, it is not a noun, or a thing to be discovered; rather, it is an act, assigning people to categories. "Oh, sure," you're thinking, "but certainly race is more natural, or more real, than, say, the category of Boston Red Sox fan." Not quite; race is intended specifically to identify a naturalistic category of people, and that is precisely where it fails.

9

Biological anthropologists for the most part don't even refer to race anymore by that word; we call it "human variation" or "human diversity." The reason is that race is a poor empirical description of the patterns of difference that we encounter within our species. That is, if by "race" we agree to denote a fairly large, fairly homogeneous, and fairly discrete—in other words, a fairly "natural"—division of the human species.

Rather, the pattern that we find is that, first, human groups tend to be distinguished culturally, rather than biologically. The differences that are generally the most important are those of dress, speech, taboo, etiquette, values, and the like—what we call "culture." The people who hate each other the most are only rarely the most biologically distinct; they are simply the worst neighbors.

Second, human groups blend into one another. As anthropologists have long known, the principal reason is that your neighbors may be loathsome, but they are the ones you trade with, and whose daughters you take in marriage (or for less lofty purposes). The nice name for this is gene flow, and its ultimate effect is to make your own descendants' gene pool that much more similar to your neighbor's.

And third, at a first approximation at least, there are all kinds of people everywhere. The amount of genetic diversity within groups is measured to be about eight times greater than the amount of variation from group to group. That means that gene pools are not discrete; they overlap tremendously. For example, you can find someone with type O blood literally anywhere on earth.

Ultimately, people are different, and groups of people are different. But the pattern of those differences does not appear to be as the commonsensical view of race would have it. People are simply similar to those nearby and different from those far away, but that does not tell you that there are four kinds of people, or twelve kinds of people, or thirty kinds of people.

And that, in a nutshell, is what roughly two centuries of scientific study of the human species have told us.

Unfortunately, there is always more to race. After all, there's got to be a reason to classify people. Otherwise, why bother doing it? And as it turns out, the very title of this book—*Race and Reality*—helps to expose the other side of race as a long-standing tool of scientific racists—people who would compromise the good name of science in the service of ignoble political goals.

In 1961, whether black and white children should be able to go to the same schools was still an open and debated question in American society. In Atlanta, a businessman named Carleton Putnam began to panic at the rising support for racial equality in America. He sought to explain, in plain English, what was wrong with the idea of school integration. His book was called *Race and Reason* and was circulated quite widely, especially in the South. Not only did Putnam explain what was wrong with the concept of racial equality, but he also explained where it came from—it was the result of a conspiracy of commies, anthropologists, and Jews.

Suffice it to say that the work was quickly repudiated in scholarly circles. In retrospect, we can see that the people who need races the most are the racists; without the reality of races, their political platform collapses.

Putnam recognized the political tide was against him, and his voice became more shrill. In 1967, the year of *Sgt. Pepper's Lonely Hearts Club Band*, Putnam published a follow-up to his earlier tract. He explained, in even more paranoid tones, how the academy had fallen victim to the wiles of those commie Jew anthropologists. The intervening years, however, had indeed seen a change in American social attitudes, and the new book did not do very well. It was dismissed as a dying ember of the archaic social villainies that America was struggling, and will always be struggling, to transcend.

Carleton Putnam's 1967 book was also called *Race and Reality*.

I hope Guy P. Harrison's book does a lot better.

Dr. Jonathan Marks
biological anthropologist
Charlotte, North Carolina

ACKNOWLEDGMENTS

I am deeply grateful to Editor-in-Chief Steven L. Mitchell and Prometheus Books for their belief in this book. The work of editors Jade Zora Ballard and Ann O'Hear was brilliant and essential.

My wonderful family—Sheree, Natasha, Jared, and Marissa—never fail to supply the love and encouragement that mean everything to me. My friend, Ms. Andrea Roach, was a crucial source of advice and criticism.

I am indebted to the experts who provided invaluable comments and guidance. They are Dr. Jefferson M. Fish, professor emeritus of psychology at St. John's University; Dr. James Flynn, professor emeritus of political studies at the University of Otago; Dr. Nick Wynne, historian and former director of the Florida Historical Society; Jane Elliott, antiracism activist and educator; Steve Olson, science writer; Dr. Joseph L. Graves, professor of biological studies; Dr. Jonathan Marks, biological anthropologist; Ken Tanabe, antiracism activist and founder of Loving Day; Dr. Curtis W. Wienker, physical anthropologist; Dr. Richard Nesbitt, psychologist and intelligence expert; Dr. Milford M. Wolpoff, paleoanthropologist; Dr. John Hoberman, expert on racism in sports; Dr. Chris Stringer, paleoanthropologist and research leader in human origins at the Natural History Museum (London); Dr. Robert Kurzban, psychology professor and race researcher; and Mike Barrowman, Olympic champion.

I also thank the many individuals around the world who shared with me their personal thoughts and intimate memories relating to the concept of race. Their comments were an invaluable contribution to this book.

INTRODUCTION

Truth is the beginning of every good thing.
—Plato[1]

**We live in a fantasy world, a world of illusion.
The great task in life is to find reality.**
—Iris Murdoch[2]

Race belief and racism seem as if they are here to stay, in the blood and in the mind forever, a permanent fixture of humanity. But are they? I don't think so, because race belief rests upon a foundation of mistakes, lies, ignorance, pseudoscience, and delusion. This means racism may prove to be far easier to rid ourselves of than we ever imagined. The first step, however, is learning the scientific realities of our biological diversity that remain lost in the fog of race belief. My hope is that this book will contribute to a new awareness of the many problems with races as biological categories of human beings.

If you are a typical race believer, and almost everyone is to some degree, perhaps when you close this book you will see an entirely new world around you. There will still be divisions and distance between dif-

15

ferent "kinds" of people, of course; only now maybe they won't seem quite so permanent or natural. Maybe you will recognize them as the product of cultural differences rather than biological differences. You might even feel a new kinship with all people for the first time and find your worldview is changed for the better. All of these things happened to me when I investigated the concept of race. I hope your experience is similar.

It is difficult if not impossible to be consistent with labels when writing about race, especially when one is challenging the very existence of races. Rather than stubbornly attempt to make every sentence conform to my conclusions about race, I have tried to make the text as convenient for the reader as possible. My use of common race labels throughout this book to describe groups of people may strike some readers as a contradiction. (Why refer to "Asians" when the claim has been made that "Asian" is not a valid biological race?) I decided that the best way to write about race for a general audience made up mostly of race believers is to "speak their language." I also refrained from placing every race name in quotes. It was tempting to do this in order to avoid any perceived endorsement of such labels or race belief. However, I felt this might become tedious or obnoxious to readers, so I have placed race names in quotes only where I felt it was absolutely necessary.

Some important areas of interest regarding race can be complex and highly technical. However, I have done my best to make everything in this book accessible to a general audience.

Race is a controversial subject and there is little doubt that this book will draw criticism from some people who will accuse me of manipulating science and twisting reality to conform to my political views. But this is not about political correctness or any sort of political agenda, left or right. I am opposed to racism, yes. I do wish everyone in the world would get along, be fair, and live in peace. But I do not feel that I need to lie, embrace pseudoscience, or betray reason in order to promote these desires. I believe that dishonesty and malicious propaganda are unnecessary in the effort to combat racism and point out the many problems of the race concept. I oppose gender discrimination, for example, but this does not lead me to deny the existence of different genders. When it comes to challenging race belief, truth and reality do just fine.

I reject the popular belief that there are naturally occurring, discrete biological categories of humans called races. I reject this claim primarily not because I think belief in races is dangerous—which it is—but because I understand it to be false, based on scientific evidence and sound arguments from scholars. The cultural groups that we call races may be real, but they are not the biological subspecies of humanity that billions of people believe them to be. This is something people need to hear about. It is nothing less than information that could change the world for the better.

Readers who feel that their race is a vital part of their identity should know that they do not have to worry that the ideas on the following pages are meant to rob them of something they hold dear. To be clear, this book argues against the existence of *biological* races. *Cultural* races are another matter.

The primary theme of this book is that biological races are not real. Another message is that we do not need to ignore our differences or learn to tolerate one another in order to draw close and become a human family. We already are that close family. Dysfunctional and feuding though we may be, we are a family. This is not hope, dreams, or optimism. It is scientific fact. Nature did not rip us apart. We did, in our minds. It is only our perceptions and the actions we choose that lead us to feel so distant from others. We do not need to come together, only to open our eyes and see that we already are together.

Whether they admit it or not, authors love to imagine that their books will last forever, standing on library shelves through the ages like glorious Greek columns. Authors share their ideas, tell stories, and then dream of how the words might outlive them. But not this time. Nothing would please me more than to see this book become obsolete and unnecessary. I want social progress to render the words within this book outdated, comical, and strange. My hope is that, sooner rather than later, this book will become nothing more than an unusual artifact from that odd time in the past when people had not yet awakened from their race belief.

NOTES

1. Hugh Rawson and Margaret Miner, eds., *The New International Dictionary of Quotations* (New York: Signet, 1986), p. 382.

2. Ibid., p. 305.

Chapter 1

BIOLOGICAL RACES ARE NOT REAL

The idea of race in the human species serves no purpose. The structure of human populations is extremely complex and changes from area to area; there are always nuances deriving from continual migration across and within borders of every nation, which make clear distinction impossible.
—Luigi Luca Cavalli-Sforza, Francesco Cavalli-Sforza[1]

The lay concept of race does not correspond to the variation that exists in nature.
—Joseph L. Graves Jr., evolutionary biologist[2]

Race is a human invention. . . . We made it, we can unmake it.
—Evelyn Hammonds[3]

Science often violates simple common sense. Our eyes tell us that the Earth is flat, that the sun revolves around the Earth, and that we humans are not animals. But we now ignore that evidence

of our senses. We have learned that our planet is in fact round and revolves around the sun, and that humans are slightly modified chimpanzees. The reality of human races is another commonsense "truth" destined to follow the flat Earth into oblivion.

—Jared Diamond[4]

Our evolutionary history is a continuous process of combining the new with the old, and the end result is a mosaic that is modified with each birth and death. This is why the process of using genetics to define "race" is like slicing soup: "You can cut wherever you want, but the soup stays mixed."

—Charles N. Rotimi, director of the Center for Research on Genomics and Global Health[5]

What you make of race depends on what the question is. And who wants to know.

—Peter Smouse[6]

Human history becomes more and more a race between education and catastrophe.

—H. G. Wells

Few things are more real than races in the minds of most people. We are different. Anyone can see that. Look at a "black" person and look at an "Asian" person. If a black Kenyan stands next to a white guy from Finland we all can see that they are not the same kinds of people. Obviously they belong to different groups and these groups are called races, right?

Well, not exactly. We have biological diversity, of course. Yes, in many cases—not all by any means—we can say something about people's characteristics based on how they look. We can see obvious differences between people who have relatively recent ancestors from different continents. None of this, however, is enough to allow us to conclude that the idea of biological races makes sense. In a nutshell, some key problems are as follows:

- Even people who look very different are not as different as you may imagine. We are a very closely related species. Also, the true ancestry and genetic makeup of a person is not necessarily revealed by superficial traits such as skin color and hair type;
- Observable features such as skin color, nose shape, and hair type are spread across humankind in a manner that is too inconsistent for us to sensibly use them as markers for separating people into groups;
- Races have been defined and used in different ways by different cultures and in different time periods. There has never been a consistent, uniform vision of the concept of race;
- People have always mated frequently and with great enthusiasm across whatever cultural borders and artificial barriers have been placed in their way. Simply put, the human story defies any attempt to sensibly divide up all people into a few discrete categories with meaningful boundaries around them.

It is clear that the majority of the world's people have not received the memo, but many scientists have been saying for decades now that biological races are not real. They do not exist. They do not occur naturally. We made them up. Generations after generations of children have been taught to believe in races as if their existence is a matter of commonsense knowledge and something nature forced on us. Meanwhile, scientists keep saying that we are one species and that's it. We are not a collection of biological races or subspecies. Races, they say, are cultural creations rather than biological realities. There are no walls between us except those we imagine.

The billions of humans alive today simply do not fit into neat and tidy biological boxes called races. Science has proven this conclusively. Humankind has always mated and migrated with no regard for the convenience or desires of taxonomists and racists. Yes, some genetic traits may cluster across the population in ways that represent prehistoric groups that adapted to different environments. But no meaningful and consistent boundaries can be drawn around these clusters to identify traditional racial groups, certainly not in a manner that conforms to the typical layperson's idea of race. For example, most Americans think of all sub–Saharan

Africans simply as "black people." But this ignores the complex reality of who they really are genetically and how their kinships accurately sort out on that continent as well as globally.

People are born every day and people die every day. Every time these things happen, the deck is shuffled anew. Humankind is a not a snapshot. It's more like a film—on fast-forward. If it were really important, and you just had to draw borders around people to divide humanity up into some sort of racial-genetic-ancestral groups, then your effort would be outdated and invalid by the next morning. The concept of race, as most people today think of it, is not scientific and goes against what is known about our ever-changing and complex biological diversity. In light of this, the toll of racism and the immeasurable lost potential of our species due to race belief are tragedies made even more terrible because they are based in fantasy.

PEOPLE WHO KNOW PEOPLE DON'T BELIEVE IN RACE

An important thing to know about the concept of race is that it is rejected by most of the people who know the most about human diversity. Anthropology is the study of humankind, our origins, our culture, our biology, our diversity—and our races if they were real. Anthropologists in previous centuries believed in races and worked very hard to illuminate race categories and expand knowledge about our racial differences. In the end, however, they couldn't do it. So the field had no choice but to shift toward the position of no races. Today's anthropologists are men and women who have looked at the question of race longer and harder than experts in any other scientific discipline. If biological races existed, anthropologists would have found them by now. But they haven't. There is still some debate, but the majority of today's professional anthropologists dismiss the idea of races as reasonable units for organizing humankind.[7] Don't misunderstand; I would never suggest that we should bow blindly to their authority and believe or disbelieve in whatever they say simply because they say it. Anthropologists have been wrong about race before and they could be wrong about race again. It is the quality of their evidence and the power of their arguments, however, that demand our attention.

In 2007, the University of New Mexico hosted a symposium entitled "Race Reconciled? How Biological Anthropologists View Human Variation." It was attended by scientists specializing in genetics, forensics, bioarchaeology, paleoanthropology, and human biology. In 2009, the *American Journal of Physical Anthropology* published a special issue on the symposium that contained several papers about race. The introduction to that issue listed five key points of agreement among the scientists. One of them in particular caught my eye: "Race is not an accurate or productive way to describe human biological variation."[8]

Anthropologists Heather J. H. Edgar and Keith L. Hunley also wrote the following in that introductory paper: "Specialists in informal education talk about 'naive notions,' which, in the context of education in biological anthropology, are the ideas our students have when they walk in the doors of our classrooms. Often, these ideas are typological, even when they are not racist. Although we have now been teaching for generations that races do not exist, these naive notions persist and they continue to have social and scientific consequences. This may be because we have failed to offer a clear and satisfactory explanation that meshes with students' lived experience."[9]

In 1998, the American Anthropological Association (AAA) issued a strong statement on race. The following are a few key excerpts from it:

> In the United States both scholars and the general public have been conditioned to viewing human races as natural and separate divisions within the human species based on visible physical differences. With the vast expansion of scientific knowledge in this century, however, it has become clear that human populations are not unambiguous, clearly demarcated, biologically distinct groups. Evidence from the analysis of genetics (e.g., DNA) indicates that most physical variation, about 94%, lies within so-called racial groups. Conventional geographic "racial" groupings differ from one another only in about 6% of their genes. This means that there is greater variation within "racial" groups than between them. In neighboring populations there is much overlapping of genes and their phenotypic (physical) expressions. Throughout history whenever different groups have come into contact, they have interbred. The continued sharing of genetic materials has maintained all of humankind as a single species.

> Historical research has shown that the idea of "race" has always carried more meanings than mere physical differences; indeed, physical variations in the human species have no meaning except the social ones that humans put on them. Today scholars in many fields argue that "race" as

it is understood in the United States of America was a social mechanism invented during the 18th century to refer to those populations brought together in colonial America: the English and other European settlers, the conquered Indian peoples, and those peoples of Africa brought in to provide slave labor.

"Race" thus evolved as a worldview, a body of prejudgments that distorts our ideas about human differences and group behavior. Racial beliefs constitute myths about the diversity in the human species and about the abilities and behavior of people homogenized into "racial" categories. The myths fused behavior and physical features together in the public mind, impeding our comprehension of both biological variations and cultural behavior, implying that both are genetically determined.

Racial myths bear no relationship to the reality of human capabilities or behavior. Scientists today find that reliance on such folk beliefs about human differences in research has led to countless errors.

We now understand that human cultural behavior is learned, conditioned into infants beginning at birth, and always subject to modification. No human is born with a built-in culture or language. Our temperaments, dispositions, and personalities, regardless of genetic propensities, are developed within sets of meanings and values that we call "culture." Studies of infant and early childhood learning and behavior attest to the reality of our cultures in forming who we are.[10]

The AAA's statement on race is eloquent and enlightening. It has a big problem, however. Nobody read it. Not enough people, certainly. It has not trickled down to the vast majority of people. In all the years since the release of these conclusions in 1998, I have not encountered one person familiar with them who was not an anthropology student or in some way seriously involved in the study of race. The public doesn't have philosophical or technical disagreements with the AAA's statement. People can't because they have never read it! This is a significant problem. At the very least, the public should know that the idea of races is widely rejected within the scientific community—at least among most anthropologists, the people who know the most about people—and that modern science has overturned many of the traditional assumptions that have propped up race belief generation after generation.

Anthropologist Robert W. Sussman, a former editor-in-chief of the *American Anthropologist*, the flagship journal of the AAA, suggests that it will take a lot more than a statement to get members of the public to pay attention. "The folk concept of race in America is so ingrained as being biologically based and scientific that it is difficult for people to see otherwise," he said.[11]

Two years before the AAA's statement on race came out, the American Association of Physical Anthropologists published its own sweeping statement on race. Keep in mind that physical anthropologists, or biological anthropologists as they are also called, are the experts most directly involved with the many aspects of human biological diversity. Some key points made in the statement are as follows:

All humans living today belong to a single species, *Homo sapiens*, and share a common descent. Although there are differences of opinion regarding how and where different human groups diverged or fused to form new ones from a common ancestral group, all living populations in each of the earth's geographic areas have evolved from that ancestral group over the same amount of time. Much of the biological variation among populations involves modest degrees of variation in the frequency of shared traits.

There is great genetic diversity within all human populations. Pure races, in the sense of genetically homogenous populations, do not exist in the human species today.

There are obvious physical differences between populations living in different geographic areas of the world. Some of these differences are strongly inherited and others, such as body size and shape, are strongly influenced by nutrition, way of life, and other aspects of the environment. Genetic differences between populations commonly consist of differences in the frequencies of all inherited traits, including those that are environmentally malleable.

For centuries, scholars have sought to comprehend patterns in nature by classifying living things. The only living species in the human family, *Homo sapiens*, has become a highly diversified global array of populations. The geographic pattern of genetic variation within this array is complex, and presents no major discontinuity. Humanity cannot be classified into discrete geographic categories with absolute boundaries. Furthermore, the complexities of human history make it difficult to determine the position of certain groups in classifications. Multiplying subcategories cannot correct the inadequacies of these classifications.

Generally, the traits used to characterize a population [skin color, hair type, and facial features, for example] are either independently inherited or show only varying degrees of association with one another within each population. Therefore, the combination of these traits in an individual very commonly deviates from the average combination in the population. This fact renders untenable the idea of discrete races made up chiefly of typical representatives.

The human features which have universal biological value for the survival of the species are not known to occur more frequently in one population than in any other. Therefore it is meaningless from the biological point of view to attribute a general inferiority or superiority to this or to that race.

The human species has a past rich in migration, in territorial expansions, and in contractions. As a consequence, we are adapted to many of the earth's environments in general, but to none in particular. For many millennia, human progress in any field has been based on culture and not on genetic improvement.

Mating between members of different human groups tends to diminish differences between groups, and has played a very important role in human history. Wherever different human populations have come in contact, such matings have taken place. Obstacles to such interaction have been social and cultural, not biological.

Partly as a result of gene flow, the hereditary characteristics of human populations are in a state of perpetual flux. Distinctive local populations are continually coming into and passing out of existence. Such populations do not correspond to breeds of domestic animals, which have been produced by artificial selection over many generations for specific human purposes.

The biological consequences of mating depend only on the individual genetic makeup of the couple, and not on their racial classifications. Therefore, no biological justification exists for restricting intermarriage between persons of different racial classifications.

There is no necessary concordance between biological characteristics and culturally defined groups. On every continent, there are diverse populations that differ in language, economy, and culture. There is no national, religious, linguistic or cultural group or economic class that constitutes a race. However, human beings who speak the same language and share the same culture frequently select each other as mates, with the result that there is often some degree of correspondence between the distribution of physical traits on the one hand and that of linguistic and cultural traits on the other. But there is no causal linkage between these physical and behavioral traits, and therefore it is not justifiable to attribute cultural characteristics to genetic inheritance.

The genetic capacity for intellectual development is one of the biological traits of our species essential for its survival. This genetic capacity is known to differ among individuals. The peoples of the world today appear to possess equal biological potential for assimilating any human culture. Racist political doctrines find no foundation in scientific knowledge concerning modern or past human populations.[12]

WHY WASN'T I TOLD ABOUT THIS?

One day in the 1980s, I sat in the front row in my first undergraduate anthropology class, eager to learn more about this bizarre and fascinating species I was born into. But I got more than I expected that day as I heard for the first time that biological races are not real. After hearing several perfectly sensible reasons why vast biological categories don't work very well, I started to feel betrayed by my society. "Why am I just hearing this now?" I thought to myself. "Why didn't somebody tell me this in elementary school? Why was I reading *To Kill a Mockingbird* in high school instead of *Man's Most Dangerous Myth: The Fallacy of Race*, by Ashley Montagu?

No races? This is big news! Why isn't it on the front page of every newspaper in the world? People have been hating and killing each other all this time over something that's not even real? Doesn't somebody think it might be important to share this information? But then I drifted into doubt. Hold on, I thought, maybe this is just a scam to get people to stop being racists. More propaganda from The Man. Maybe the wizards behind the curtain are just saying this stuff hoping to get us, the idiotic public, to stop spending so much time and energy on racism so that we can focus more on watching television commercials and buying consumer goods. Besides, race *has* to be real. People really do look different. They come from different places originally. A lot of black guys are in the NBA and a lot of white guys are in the US Congress. Don't people of different races get different diseases? They live in different neighborhoods. They dance differently. Race *must* be real.

I was skeptical but kept listening, kept reading, and kept thinking, nonetheless. The result of my efforts was that I came to the inescapable conclusion that the races I had been taught to believe in really were cultural fantasies masquerading as biological reality. A year later, I had the good fortune to take Dr. Curtis W. Wienker's paleoanthropology class. Dr. Wienker is a physical or biological anthropologist, an expert on early humans and biological diversity. He has sometimes worked with law enforcement as a forensics consultant. By this time, I was a hardcore anthropology and history student with an insatiable hunger for knowledge. I wanted to know everything possible about early hominids, cultures different from mine, and, of course, race. Those were heady days, slightly embarrassing to me now, when I could literally lose sleep wondering about *Homo erectus* taming fire or the sounds of Neanderthal conversations.

Dr. Wienker was an excellent teacher. He was not exactly the warm and cuddly type, but his lectures were so loaded with fascinating information that I always looked forward to his classes. He also made the effort to bring in interesting people for guest lectures. The impressive guest list during my semester included Donald Johanson, the discoverer of the famed Lucy fossils, and Vincent Sarich, co-discoverer of the molecular clock (the use of DNA to determine when two species diverged). Probably more than anyone, Dr. Wienker enlightened me on the idea that races are just not real in the way most people think they are. I tracked him down while conducting research for this book and asked him if anything had come up in the last twenty years that might have given him reason to doubt his position on races. Not even close, he says. He is still confident that while our biological diversity is very real, biological races are not. The biggest mistake people make in thinking about race, Dr. Wienker says, is that they "think in terms of categories and types, which obscures the amazing, manifest biological variation within human populations." Humankind, he says, doesn't come in pieces. It's one big beautiful blended gob of genetics and tangled ancestry.[13]

"The use of race as a convenient mechanism by which humans might be classified persists for several reasons," said Dr. Wienker. "Most are just not aware of the fact that it is a social construct lacking reality or biological validity. And it has been used for so long that, like such things, it is very slow to be changed. As well, anthropologists have not done as good a job as they might about educating others as to their scientific posture with respect to human biological variation."[14]

Dr. Wienker also sent me this straightforward statement on the concept of race. It is so well written that I dare not paraphrase his words:

> The reason the biological concept of races is not valid and is of little, if any, use when applied to modern humanity is simple. It is not possible, with reproducible consistency acceptable to scientists, to classify even a majority of humans into an agreed upon finite number of races. Our biological variation is simply not consistent across space, partly because of prehistoric and historic migrations. Such migrations were often culturally mediated through such institutions as slavery and conquest; they resulted in considerable admixture, often between groups that were quite different in their patterns of biological variation. These circumstances spread far back into our past and because they have been taking place for so many millennia, the net result is an extremely muddled pattern of human biological diversity across geographical space today. We in the United States are just unable to see and grasp it because the biological variation observable to us in our everyday lives is relatively finite, at least in most cities.[15]

Finally, Dr. Wienker adds that he is concerned about the challenges of teaching university students the truth about race and the realities of our biological diversity, because their perception of race is firmly entrenched by that stage of their lives: "By the time students reach university classrooms, the traditional view of the reality of human races has become indelibly ensconced in their cognitive maps, often irrevocably so. Thus, education as to the reality of human races as social constructs having no biological validity needs to occur earlier in the developmental cycle and educational process. Units on the reality of human races as social constructs, and not as biologically real categories, should be directed at preadolescents whose cognitive maps have not yet become lithified."[16]

I'll vouch for that. When I heard professors lecturing about the problems with race for the first time in college, it was something of a struggle to accept their ideas and evidence. It felt wrong and nonsensical at first glance. After a childhood of total immersion in race belief, reversing my thinking presented a challenge. I only knew two ways: be a racist and hate people based on race, or be nice and try to love all people, even those who belong to races that are very different from yours. Both options assumed that races were real. I was intrigued by the claim that races are in the mind and not in the blood. But I was skeptical, mostly out of loyalty to everything I knew. It was a slap in the face, like being told that mom, dad, and the good old United States of America had played a prank on me...like learning that my parents had lied to me about Santa Claus.

At the very least, young children should be exposed to the current anthropological view of race. This is the branch of science that is directly dedicated to the study of human culture and biology. We ignore the experts in this branch of science at our collective peril. Racism is such a wasteful and destructive force that this is nothing less than a matter of life and death. Failing to enlighten children about the realities of race seems to me like educational malpractice. I never should have made it through twelve years of schooling, before entering a university, without ever hearing the important news that most anthropologists reject the concept of biological races.

While writing this book, I communicated with Milford M. Wolpoff, one of the world's leading anthropologists. He sent me one of his papers, which included this statement: "The idea that there were once pure human races is dead and buried, and if race cannot reflect common descent, and if there is no validity to the precept that human races are constellations of biological characters that show greater differences between each other than variation within one of them, race can only have a social definition. There simply are no clearly distinct types of humanity, and there is no racial taxonomy for the living."[17]

Don't you think these are important words, relevant ideas, that children should be exposed to early in life? This is not propaganda or dogma. It is basic science. We teach children that the Earth is round and revolves around the Sun. Why not "There simply are no clearly distinct types of humanity," too?

Dr. Nick Wynne, an American historian and a former director of the Florida Historical Society says:

I think this idea should be broached in public education systems at a very early age, perhaps even kindergarten. Of course, the problem comes in trying to get such ideas accepted as part of the day-to-day curricula. As Theodore Roosevelt once remarked, "The most independent political bodies in the United States are local school boards." While the acceptance of federal funds has somewhat lessened their independence, school boards at the local and state level are generally not composed of professional educators and tend to bring their personal attitudes into office with them. Even 'professional' educators are frequently little more than average people, armed with a college degree, that view the educational system as primarily providing a job and secondarily as providing a forum for inculcating their own ideas. Yes, such ideas should be taught, but the critical question remains as to how these ideas would be incorporated into the educational systems.[18]

California State University, Los Angeles, sociologist Yehudi Webster believes that America's current educational system actively promotes race belief, to the detriment of the nation. He feels the fact that classrooms are "devoted to producing workers and consumers rather than cultivating human minds in a spirit of critical thinking provides the basis for the acceptance of the fallacies in racial classification. In this sense, America has not racial problems but philosophical educational deficiencies that manifest themselves in the racialization of social relations."[19]

"Education is the only way to reduce racism," said Steve Olson, author of *Mapping Human History*, an outstanding book about biological diversity, "along with direct experiences with people of different 'races.' I was quite confused myself before beginning *Mapping Human History* about the relations among human groups. While some aspects of those relations remain unknown—and may never be known—enough is known to dispel almost all of the misconceptions most people have about race. That's one reason why I've always been so eager to talk about these subjects—you can practically see people in the audience getting it as you go along."[20]

TRUTH ABOVE POLITICAL CORRECTNESS

During the process of writing this book, I corresponded with a well-known scientist in the hope that he would share his thoughts about race for the book. Ultimately he declined, apparently because he disagreed with my position that biological races are cultural creations. I accepted his choice, of course, but what disappointed me is that during the course of our exchanges, his comments morphed from those belonging to a scientific discussion about the reality of races to those belonging to a discussion about conservatives versus liberals. To be clear, I view this as an issue about what is real and what is not, one that is far too important for us to center on tribal political philosophies that are ultimately irrelevant. This book is not pushing any political agenda—unless reality can be considered a political agenda. I want liberals, conservatives, and everybody else to understand what is true and what is false about our biological diversity. Yes, some issues associated with race and racism are undoubtedly political and can be viewed as reflecting liberal or conservative views, but those issues are secondary to the question of whether or not biological races exist.

HOW MANY OCEANS?

I give science and history lectures at the elementary school my children attend. It's a great part-time job. I do earn a salary, but it's so enjoyable that I would do it for free. I get to be around my own children in their school environment, and it's nice to feel I'm contributing something meaningful to the world. It also gives me the chance to finally show off my museum-quality replica hoplite helmets and hominid skulls to people who actually find them interesting.

In order to present lectures that are both engaging and thought provoking, I try to simplify complex ideas and explain them in ways that make them accessible to young children. I've hit my young audiences with many heavy topics and they haven't rebelled yet. For example, the day after a lecture on the universe, a frantic student approached me and confessed that she just wasn't able to "completely understand" dark matter and dark energy. She relaxed noticeably when I explained to her that the world's top scientists have not yet figured those things out either. I was pleased to see my talk had not left her intimidated or overwhelmed. She had gone home and attempted to singlehandedly solve the mysteries of dark matter and dark energy! For the purposes of this book, I attempted to come up with a way to talk about the concept of race that might help people of all ages and educational levels grasp the problems with it.

My first example is quick and easy. Imagine six gigantic boxes in which we can place all the people in the world. Let's sort out everyone into six groups: "Asians," "blacks," "whites," "Latinos," "Native Americans," and "Pacific Islanders." But wait, there is something very unusual about our boxes. They don't do a very good job of keeping the different kinds of people apart, because these bizarre containers don't have tops, bottoms, or sides. We know the boxes are there because we say they are there, so we keep sorting people into them. But the fact is, nobody can really see or feel the sides, tops, or bottoms of them. Meanwhile, millions of people keep moving from box to box, pretty much as they please. You might be wondering what significance these boxes have if we can't see them and they are unable to contain the people we put into them. Do these boxes really exist, or is there some sort of "emperor's new clothes" scenario going on? This is what biological races are: boxes without tops, bottoms, or sides— inadequate, to say the least.

Here's a better one. If I asked you how many oceans there are, you would probably say three, four, or five, depending on how well you did in geography class. Most people get Pacific, Atlantic, and Indian easily. Many also name the Arctic Ocean and a few might remember to include the Southern Ocean. The correct answer is "five oceans."

But wait, that's not the answer that matches the reality we find in the world. Five is the correct *cultural* answer. *Nature's* answer is "one." That's right, there is just one continuous body of salt water that flows around our planet. It is not divided into five oceans by nature. We did that. If you doubt it, simply look at a globe and you will see that the water is connected. Trace your finger over the "oceans" to discover the reality of our planet's single ocean. Flip the globe upside down for the best view of how connected and free flowing the water is.

It is only tradition and shortsightedness that leads us to think there are multiple distinct oceans. One can understand the need to give names to regions or points of reference on a map, but believing in isolated oceans is incorrect and very misleading about the real nature of our world's salt water. It is also dangerous for us to think of separate oceans, given the problems of pollution and loss of biodiversity, which are global, one-ocean problems. It is because people long ago saw not one but many oceans that we "see" five different oceans today. We gave them names and children were taught to believe in them, generation after generation. Sound familiar?

Our species is a lot like the salt water of our planet. Just because we named the oceans, that doesn't mean they are separate in reality. It doesn't matter how many people obsess over and measure differences between the area of water we call the Indian Ocean and the area of water we call the Pacific Ocean. They still belong to the same larger body of water. Likewise, it doesn't matter that some people obsess over our differences while ignoring our common ancestral origins and continuous mixing of genes. We all still belong to the same larger species. If people want to argue that the ideas of multiple oceans and multiple races are not made up, simply ask them to show you exactly where the borders are. Where does one ocean end and the other begin? Where does one race end and another begin? They can't do it because these divisions do not exist in nature. Look at a globe, travel the world. Look beyond the labels and you will discover one ocean and one human species.

NOT BELIEVING IN BIOLOGICAL RACES DOES NOT NECESSARILY MEAN LETTING GO OF RACE

Exploring our biological diversity is fascinating and important, but it can also be confusing and frustrating. For example, one cannot simply accept the race-is-not-real position and call it a day. It's not that easy. Race *is* real. Races may not exist as the biological categories many people imagine them to be, but they are still here with us, nonetheless. Races are real because we made them real. If we tossed out all uses of race today, for example, we would find it very difficult to study and find solutions to the serious problem of racial disparities in health and healthcare, for example. If some people are experiencing racism, facing limited opportunities, and living in inferior or dangerous environments due in some part to their membership in a socially constructed race group, then race can't be ignored just yet. There is no contradiction here. *Biological* races are not real; *socially constructed* races are very real.

The American Sociological Association, an organization that represents some thirteen thousand sociologists, issued a statement on race in 2003 that sought to justify their continued use of "race" in sociological research. While acknowledging the fact that "biological research now suggests that the substantial overlap among any and all biological categories of race undermines the utility of the concept for scientific work in this field," the statement declared:

> [M]easuring differential experiences, treatment, and outcomes across racial categories is necessary to track disparities and to inform policy-making to achieve greater social justice.... Sociological scholarship on 'race' provides scientific evidence in the current scientific and civic debate over the social consequences of the existing categorizations and perceptions of race; allows scholars to document how race shapes social ranking, access to resources, and life experiences; and advances understanding of this important dimension of social life, which in turn advances social justice. Refusing to acknowledge the fact of racial classification, feelings, and actions, and refusing to measure their consequences will not eliminate racial inequalities. At best, it will preserve the status quo.[21]

This, I feel, is a sensible position. Mythological though race may be, belief in it causes real-world problems that most likely, to be successfully analyzed, require the use of social "race" categories. Collecting data and researching "race" may appear to be negative because it perpetuates race belief. But it may be necessary in order to generate needed information and aid social progress. However, it is vital that scholars and governments actively promote a reality-based view of what races are whenever they use racial categories. When sociologists write about races, it would be helpful if they would consistently include disclaimers, discouraging the belief that races are discrete biological subsets of humanity. Races are socially constructed groups, and scholars have an obligation to spread this important knowledge widely.

It also is important to be aware that cultural races can and do have profound biological influences on people. For example, a long-term study published in 2009 found that the stress of poverty in childhood could cause short-term memory impairment later in adulthood. This is an obvious problem that could negatively impact education and employment. Given the fact that some races in many societies are significantly overrepresented among the impoverished, it is easy to see how membership in a socially constructed race may carry serious biological implications.[22]

Finally, if people enjoy some psychological benefit from their identification with a race group, then keep it, I say. While it can be argued that a raceless world might lead to a reduction in negative thought and behavior (although surely not to the elimination of all prejudice, of course), it can also be argued that there is nothing wrong with embracing and maintaining an identity tied to a racial group, as long as one doesn't buy into all the irrational beliefs about genetic limitations that surround race belief. For example, there is nothing necessarily wrong with black kids in America dreaming of becoming professional basketball players one day because they think it's cool and it's something black people excel at. There is something very wrong, however, if they think "black people" are good basketball players only because they have African genes and not because they work so hard at it. Even worse would be to believe that a career in physics or chemistry is not a realistic goal, because "black people" are not genetically predisposed to excel in those fields. I am suggesting that, given the importance of racial identity to some people, maybe there is a way to keep the good and let go of the bad. It is a possibility. Race may not have to be an all-or-nothing proposition.

If individuals can make the distinction between the genetic determiners of destiny that biological races have traditionally been thought to be and the socially constructed groups that races really are, then I don't see a problem with such individuals holding on to their "race" if that is what they feel they want to do. It would be so easy to simply declare that races are fabricated, dangerous nonsense that must be abandoned immediately by everyone. But for many people, I suspect, that would be like ripping out a part of themselves. Understandably, that's a little too much to ask.

WHERE DID RACE COME FROM?

Because it is tied to deep notions of biology, history, and origins (both secular and religious), many people may incorrectly believe that the concept of race has "always" been with us. Ancient humans understood human races in the same way that they understood tides, seasons, and the wildlife around them, right? No, not even close. While many ancient and prehistoric societies undoubtedly noticed and noted physical differences between some groups they encountered, there has never been a consistent view of race that is anything like the one that is currently popular in, say, contemporary America. Races are cultural creations, and as such they have something in common with cultures: they change. Even within the United

States, in just the last few centuries, race has changed dramatically. "White people," for example, didn't always include eastern and southern Europeans. Even Irish immigrants were not quite "white" when they first arrived in the New World. Race as a descriptive and organizing idea was fluid and flexible. It remains so today. For something that is supposed to be natural and based in common sense, race is oddly inconsistent.

According to Dr. Wienker, Carl Linnaeus, the eighteenth-century Swedish botanist who laid the foundation for the scientific classification of life, was the first scientist to organize humanity into different types. His "races" included Africanus, Americanus, Asiaticus, Europeanus, and Monstrosus. (I'm guessing members of the "Monstrosus" race might have had discrimination problems.) Dr. Wienker says these races were stereotyped by appearance and behavior. "For example, [Linnaeus] described the European type as fair and brawny, blue eyed, inventive, and governed by laws; the Asian type was sooty, black eyed, melancholic, haughty, and governed by opinion."[23]

The German Johann Blumenbach, another eighteenth-century scientist, famously came up with the name "Caucasian" to identify white Europeans, because he thought a skull from the Caucasus Mountains of central Eurasia was an ideal and beautiful specimen. Blumenbach wrote *On the Natural Varieties of Mankind*, a book that would prove to be very influential for a long time. In the book's second edition, Blumenbach identified five categories of people: Mongolian, the yellow race; Malayan, the brown race; Ethiopian, the black race; Caucasian, the white race; and American, the red race. Although he gave geographical names to his categories, his system marked a move away from geography toward physical appearance that would set the tone for future racial thought.[24]

"Historians have now shown that between the 16th and the 18th centuries, race was a folk idea in the English language; it was a general categorizing term, similar to and interchangeable with such terms as type, kind, sort, breed, and even species," write Audrey Smedley and Brian D. Smedley in the *American Psychologist*:

Toward the end of the 17th century, race gradually emerged as a term referring to those populations then interacting in North America—Europeans, Africans, and Native Americans (Indians). In the early 18th century, usage of the term increased in the written record, and it began to become standardized and uniform. By the Revolutionary era, race was widely used, and its meaning had solidified as a reference for social categories of Indians, Blacks, and Whites. More than that, race signified a new ideology about human differences and a new way of structuring society that had not existed before in human history. The fabrication of a new type of categorization for humanity was needed because the leaders of the American colonies at the turn of the 18th century had deliberately selected Africans to be permanent slaves. In an era when the dominant political philosophy was equality, civil rights, democracy, justice, and freedom for all human beings, the only way Christians could justify slavery was to demote Africans to nonhuman status. The humanity of the Africans was debated throughout the 19th century, with many holding the view that Africans were created separately from other, more human, beings.[25]

Dr. Wienker adds that anthropology did not have any cause for pride with regard to biological diversity research during the nineteenth century and the first half of the twentieth century: "The scientific literature on human races was filled with [work that was] subtly, and sometimes overtly, racist. Populations were viewed in terms of their external phenotypic characteristics; arguments involved how many races existed, and how they were related to one another. There was no interest in the significance of the biological variation within *Homo sapiens*. Racial characteristics and human races themselves were viewed as static entities. Primary emphasis was put on norms and modes, usually of externally visible physical characteristics—on types; individual biological variation within groups was not considered, nor typically was biochemical [nor] serological variation. Biological variation within a population, manifest in any human group, was ignored."[26]

The work of the late anthropologist Ashley Montagu was a key factor in getting his field to rethink human biological diversity. His groundbreaking book, *Man's Most Dangerous Myth: The Fallacy of Race*, was first published in 1942. It was way ahead of its time. So far ahead, in fact, that the world still has not caught up to it.

"The classificatory definition of humanity as *Homo sapiens*, properly interpreted, is appropriate because it gives quite accurate status to a unique class of creatures—human beings—characterized by an educability, a capacity for wisdom and intelligence approached by no other creature," wrote Montagu. "These traits, *not* the external physical traits, constitute the principal, the distinctive qualities, that make *Homo sapiens* human.

When human beings are defined on the basis of the differences in physical traits we narrow the definition of their humanity. And that is, perhaps, the most telling criticism of the concept of race."[27] Montagu continues:

> No one ever asks whether there are mental and temperamental differences between white, black, or brown horses—such a question would seem rather silly.

> Humans impose on nature the limitations of their own minds and identify their views with reality itself. Pixies, ghosts, satyrs, Aryans, and the popular conception of race represent real enough notions, but they have their origin in traditional stories, myths or imagination.[28]

Another book, *Races: A Study of the Problems of Race Formation in Man*, by Carleton S. Coon, Stanley M. Garn, and Joseph B. Birdsell, was an important step in anthropology's move away from race belief toward race rejection in the second half of the twentieth century. According to Dr. Wienker, these men "infused evolutionary thinking into the consideration of human races by biological anthropologists in the early 1950s. While their work contained the traditional classification of human races, it was the first to view biological variations and human populations as dynamic entities. It was also the first to consider human races in terms of the evolutionary significance of global patterns of biological variation."[29]

Most biological anthropologists and other anthropologists now agree that the biological concept of race has "little if any scientific validity and use when applied to modern *Homo sapiens*," says Dr. Wienker.[30] Nonetheless, most people today learn early in life that biological categories of humans called "races" are real and that they have a lot to do with how societies are structured. People are taught, formally or informally, that these "natural" groups determine or at least influence what people can do and how well they can do it. This worldview is reinforced continually throughout life. Race never goes away. It jumps out at us from our history books. It visits us in news stories, in sports, in music, in our friendships and working relationships. It even plays a primary role in our choices of mates and marriage partners. Race saturates virtually every aspect of our lives. Just because something may be unscientific, that does not mean it doesn't have an impact on the world. Astrology is not scientific, for example, but many people organize their lives around its claims, nonetheless. Illusions can become real factors in the real world when we give them life by believing in them. No matter how flimsy or nonexistent the scientific credibility of racial groups may be, they matter and there is a reality about them because some people use them to hurt or favor others. People have killed and died for races. You can't get any more real than that. Think of ancient Greece. Zeus, Apollo, and the rest of their gods may not have been real but the religion of the ancient Greeks was very real. It influenced the Greeks' lives, their history.

The negative repercussions of race belief lead many anthropologists to hope for a mass shift in popular thinking about race. They want people to understand that races are based on incorrect ideas about our evolutionary past and modern biology. The traditional view of race, they say, is a mistake, because biological racial groups are not isolated, consistent, or meaningful categories of people. In reality, we are a fluid, ever-changing population that defies separation and confinement by the walls we imagine to stand between us. Races can never hold up to close examination. Most Americans and Europeans today, for example, see the "black race" as a category of people in or from Africa, people who are united by their common genetics. But this is not as obvious or sensible as it may seem, because Africans are the most genetically diverse people in all the world. There are "black Africans," for example, who exhibit far greater genetic distance between themselves and other "black Africans" than they do when they are compared to some "white Europeans."

If the world's people will not or cannot wholly reject belief in racial groups, we should at least understand the truth of the biological diversity that races are supposed to be based on. Even this step, I believe, can help to move us forward, away from the worst effects of racism that race belief creates. The combination of cultural diversity that magnifies real biological differences in our minds, errors made by scientists in the past, and perhaps a natural tendency to seek patterns and think in terms of "us" and "them" has led to the deeply entrenched notion that races are an obvious reality. Although it is supposed to be obvious, however, an incredible amount of energy has been spent attempting to prove conclusively that biological races are real and relevant to society. The fact that no scientist to date has found this conclusive proof is itself a powerful statement about the problem with race. In *Mapping Human History*, Olson writes:

Past efforts by European and American scholars to draw rigid distinctions among human groups would be comical if they weren't so tragic. Scientists have devoted entire lifetimes to cataloging and measuring how beings differ. They have compared skin colors, pored over the nooks and crannies of skulls, and measured the lengths of arms, legs, torsos, and genitals. For the most part this endless poking and prodding has been pointless. Scientists today know little more than what a cursory examination of the world's people reveals—that individuals differ physically in ways that correspond roughly but not perfectly with their ancestry. Nevertheless, the hypotheses developed in the course of all this speculation about human differences have profoundly affected history. Ideas that we now know to be utterly false have had an indelible effect on the world. One of these ideas is that human beings can be divided into discrete races with specific physical and often cultural characteristics.[31]

There are so many problems with the popular concept of race that it is difficult to know where to begin. Skin color is probably the most popular racial identifier. It may also be the most ridiculous way to attempt to separate and categorize human beings. Skin color doesn't come close to holding up as an effective tool for splitting humankind into races. A simple thought experiment that one of my anthropology professors presented to me goes like this: Imagine lining up every human on Earth in one very long straight line. Arrange them according to skin color, from lightest to darkest. Now, walk down the line, paying close attention to skin color, and look for a natural place to make a break between people of one color and others. You would never find such a point between two people standing in the line, because the change in skin color along the entire line would be so gradual that no natural border would ever present itself. To achieve the goal of finding a break between colors, you would have to arbitrarily make up divisions and place them wherever you wish. This is exactly what the traditional concept of race is: made-up borders, with no natural, objective, or logical basis, that we have placed between people to create specific groups.

Let's push this a bit further. Somewhere along our imaginary line we might find a dark-skinned Tanzanian standing next to a dark-skinned Australian Aboriginal person who is in turn standing next to a dark-skinned person from India. With skin colors that are virtually identical, no sensible border can be placed between them based on color. But this contradicts the "commonsense" notion of race that is popular today. Most race believers would place dark-skinned Tanzanians, Australian Aborigines, and Indians in different races. Clearly something other than skin color determines a person's admission into a specific racial category. Against all reason, however, skin color has been used as a framework for racial divisions in many societies over the last few centuries. Decisions based on skin color have hurt people and even ended lives unfairly. It is incredible that something so superficial and trivial as color—the amount of melanin in the epidermis—could ever be given so much cultural power. During a stay in Papua New Guinea, I met some people from the island of Bougainville and was amazed to see how dark their skin was. I had never seen anything like it. I grew up in the American South around many "black" people and had also traveled in Africa but had never seen anyone with skin this dark. Skin color that is different from your own or different from what you are used to seeing may be interesting at first glance, but after that, so what? It's still just skin color and an absurd way to categorize humankind. My Bougainville friends, for example, are most likely more closely related to me, a "white European-American," than they are to most dark-skinned people in Africa or America. We use biological traits such as skin color, nose dimensions, lip size, and hair texture to help in creating socially constructed races. But race is not a biological fact.

FOLK HEREDITY

"Racial classifications represent a form of folk heredity, wherein subjects are compelled to identify with one of a small number of designated human groups," declares Jonathan Marks, a biological anthropologist. "Races do not reflect large fundamental biological divisions of the human species, for the species does not, and probably never has, come packaged that way."[32]

Dr. Marks says nothing has come out of genetic research in the last several years that has caused him to reevaluate his position on race. "Geneticists should shut up about race," Marks says, "because race is not a genetical category. It is a biocultural category." He continues:

> It is not a theory of difference, it is a theory of meaningful difference. Geneticists can analyze the differences between two people, but they cannot tell you whether they are representatives of the same group or of two dif-

ferent groups. That's what race is: an answer to the question, "Are these two kinds of people, or one?" Genetics can only tell you they're different.

I say that anyone who reads the genetics literature so credulously, without appreciating the unprecedented conflicts of interests that have arisen in human genetics the last generation, doesn't know it well enough to be brandishing it at others. [The biggest mistake people make when they think about race these days is] believing geneticists when geneticists say that they have a solution to the issue of race.[33]

Let's return to the example of lining up the entire human species according to skin color and trying to locate borders. What if we did it based on a combination of traits, instead of just one? In everyday life, people usually rely on a few different physical features to match an individual with a racial group. So, what if we imagine lining up every person on Earth, based not just on skin color but also according to some relationship between hair type and facial features as well? Wouldn't that give us the categories most people think of as the traditional races? Again, not even close. There would be no point along the line that would indicate the end of one race and the beginning of another. Furthermore, we wouldn't be able to line up the people based on external appearances that coincided with their true ancestral or genetic history. For example, I know from personal experience that there are some Fijians who, based on appearance, would seem to fit nicely into the "black race," at least at first glance. There are some people in Nepal who are too similar in appearance to indigenous South Americans to be easily picked out of a lineup. Whether it's based on one observable trait or a combination of observable traits, dividing the human population into traditional racial groups only seems easy if we ignore reality and pretend that humankind is made up of extreme physical types with nobody in between. Certainly there can be relatively huge differences between extreme examples: a white Canadian and a black Kenyan. But our world is made up of many more people than white Canadians and black Kenyans. Focusing on extremes is like saying, "Look, here's a man who is seven feet tall and here is a man who is four feet tall. Clearly the human population is made up of two races—the tall race and the short race. It's obvious!" But what about all the people who are of medium height? And where exactly do we draw the boundaries to separate short from medium and medium from tall?

Biological anthropologist Vincent Sarich is one of the most prominent race believers in the scientific community. He is an accomplished scientist with impressive credentials, but he often defends biological races with a matter-of-fact tone, as if it's all just common sense. But is it?

"The most basic evidence that race exists is the fact that we can look at individuals and place them, with some appreciable degree of accuracy, into the area from which their recent ancestors derive," he wrote in the journal *Skeptic*. "The process involved is illustrated by a simple thought experiment where one imagines a random assortment of 50 modern humans and 50 chimpanzees. No one, chimp or human, would have any difficulty in reconstituting the original 50 member sets by simple inspection. But the same would be true within our species, say 50 humans from Japan, 50 from Malawi, and 50 from Norway. Again, by simple inspection, we would achieve the same 100 percent accuracy."[34]

This might sound reasonable, but it's not. The real world is not fifty people from Japan, fifty people from Malawi, and fifty people from Norway. Try this thought experiment. Take fifty people from every other country in the world, mix them in with Sarich's tiny and unrealistic hand-picked group of extremes, and then see how easily you can separate them and accurately identify everyone's recent ancestry by simple inspection. It is also fair to ask why we are trying to create these categories in the first place, and what else, other than recent ancestry, is to be tied to them.

"The concept of race didn't exist until the invention of oceangoing transport in the Renaissance," explains C. Loring Brace, a leading physical anthropologist. "It never occurred to [early land explorers such as Marco Polo] to categorize people, because they had seen everything in between. That changed when you could get into a boat, sail for months, and wind up on a different continent entirely. When you got off, boy, did everybody look different! Our traditional racial groupings aren't definitive types of people. They are simply the end points of the old mercantile trade networks."[35]

A LOT MORE THAN COLOR

The big problem with the common perception of races is that people think each race is loaded with unique abilities and limitations, as well as personality and moral tendencies. The problem is not just that biological races are wrong; it's that they are monstrously and gigantically wrong with severe implications.

"The term race implies the existence of some nontrivial underlying hereditary features shared by a group of people and not present in other groups," writes evolutionary biologist Joseph L. Graves Jr. in his ideally titled book about race, *The Emperor's New Clothes*. "Biological science has long been interested in the identification and quantification of variation within species and has developed relatively precise tools to examine the hereditary characteristics exhibited by organisms. These developments were instrumental in allowing the Western, socially constructed concept of race and the biological concepts of race to diverge. None of the physical features by which we have historically defined human races—skin color, hair type, body stature, blood groups, disease prevalence—unambiguously corresponds to the racial groups that we have constructed."[36]

Dr. Graves is also the author of *The Race Myth: Why We Pretend Race Exists in America*, a brilliant survey of the race question. Dr. Graves is an American evolutionary biologist by trade, but it is clear that his mind is not trapped inside a laboratory. He cares about the serious real-world implications of race belief, both for his country and for the world.

"We must recognize that the underlying biological diversity of the human species cannot be artificially apportioned into races, because races are simply not biologically justified," he writes. "If we can understand that all allegiance to racism is ideological, not scientific, then we may be able to silence the bigots once and for all. We may be able to construct social systems that allow all of our citizens to actualize their biological potential. If we can live up to our creed of equality for all, then maybe we will have a chance to finally actualize the true spirit of democracy and the American dream."[37]

NOBODY RECEIVED THE MEMO

Biological races are not real. Given the impact that belief in biological races has on the world, one would think that this information would be spread far and wide, familiar to everyone, whether individuals choose to accept it or not. Outside of anthropology conferences, however, the initial reaction to the no-real-races claim from many people is awkward silence or outright laughter, because the existence of races is so undeniably obvious in their minds. Most people, in my experience, have no idea that so many scientists reject the race concept.

Nearly every scientist today agrees that during the last one hundred thousand to fifty thousand years, our ancestors spread throughout Africa, into Europe, Asia, and beyond. These people lived in different regions, with different environmental and cultural influences. A result of this was the emergence of some different traits, such as dark skin and light skin, wide noses and narrow noses. During this period, brief by evolutionary biology standards, interbreeding between many of these populations continued. The result of all this is what we have today: about seven billion people with very blurry transitions between groups/populations/clines/clusters/races (pick your term). The differences in skin color, hair, and facial features that so many people fixate on today were recent and rapid additions to our diversity. The challenge for us now is to accept our biological diversity while realizing that the traditional concept of races does not adequately or honestly describe that diversity. Race probably hasn't been a valid concept for at least 30,000 years or so. That's when the last Neanderthals lived alongside anatomically modern humans in Europe. They, biological anthropologist Marks and paleoanthropologist Milford Wolpoff tell me, were probably the last human group that could sensibly have been regarded as a race or subspecies, separate from other "kinds" of humans. Race belief has been out of date for about thirty millennia.[38]

"There is no doubt about the fact of human biological diversity," explains anthropologist Stephen Molnar. "Traditional racial divisions, however, are based on a faulty perception of human differences and a lack of understanding of the causes and meanings of these differences. Like our ancestors in previous centuries, we rely on a simple visual appraisal of appearances related to size, form, and color to identify distinctions between various groups.... All of these features have led, or I should say misled, writers to draw conclusions about the relationships between groups. Lookalikes have been considered to be closely related by common descent, while dissimilar appearances have been taken as evidence of fixed racial boundaries."[39]

Dr. Molnar also believes that the perceptions of our biological diversity will continue to change as more and more genetic data come in from around the world. This too is a blow to the existence of real and meaningful racial groups, he states. "The more we learn about the variability of population groups, the more difficult it is to fix boundaries between them."[40]

Science writer Olson favors abandoning not only the popular concept of race but the word "race" as well:

> For a long time people have tried to use the physical differences among groups to divide human beings into three races, five races, thirty races, even thousands of "microraces." Many schemes have been proposed; none has worked. There are too many exceptions, too much overlap between groups. Humans just don't sort neatly into biological categories, despite all the attempts of human societies to create and enforce such distinctions. Meanwhile the word "race" has become so burdened with misconceptions, so weighed down by social baggage, that it serves no useful purpose. The sooner it can be eliminated, the better.... [W]hen we think in terms of race, we are inevitably misinterpreting an extremely complex biological reality. In fact, all people are closely related through immeasurable lines of descent that defeat any attempt to divide humans into races.[41]

While genetic research has given plenty of ammunition to the race-is-not-real scientists, it also has become a source of new confidence for those who maintain that biological races are valid. But Jared Diamond, a physiologist and evolutionary biologist, believes that all efforts to validate traditional racial groups with genetic evidence will prove as troublesome and fruitless as trying to do it with facial features and skin color was in past centuries. He feels it is a repeat of the same game, only with fancier tools this time around. "While the biologists still haven't reached agreement, some of their studies suggest that human genetic diversity may be greatest in Africa," says Dr. Diamond. "If so, the primary races of humanity may consist of several African races, plus one race to encompass all peoples of all other continents. Swedes, New Guineans, Japanese, and Navajo would then belong to the same primary race; the Khoisans of southern Africa would constitute another primary race by themselves; and African 'blacks' and Pygmies would be divided among several other primary races."[42]

This leads us to one of the fatal problems of traditional race belief. Races can be—and have been throughout history—created based on an ever-changing assortment of criteria. Skin color, nose width, folds of skin over eyelids, hair, height, blood, genes, languages, nationality, and religion have all been used to draw arbitrary lines between us that are supposed to be based on obvious biological fact. Creating races, however, regardless of how it is done, always leads to more questions. Does the race commonly referred to as "black" include the millions of very dark-skinned people who live in India, Australia, and numerous Pacific islands? If so, it means genetics has little or nothing to do with the "black race," because those people are generally not closely related to black Africans. Should dark-skinned people in India be placed in the "white race" because their hair is similar to that of "white" Europeans? Does the "Asian race" include bushmen in southern Africa who have an epicanthic fold (a skin fold of the upper eyelid) as most Japanese and Chinese people do? People think that races are an easily recognizable biological phenomenon that nature or the gods imposed upon us when, in reality, they are almost entirely creations of perception that are maintained, modified, and many times forgotten by cultures. (Yes, racial beliefs—thought to be natural, obvious, and vital—often just fade away unceremoniously as cultures change over time.) Popular ideas about what races are and common usage of the term are not only unscientific but also contradict common sense in many cases. Some people today refer to the "Jamaican race," for example. I have been to Jamaica and know many Jamaicans. Jamaicans are not a biologically homogeneous population. The Jamaican national motto, "Out of many, one people," provides an obvious clue to their diverse ancestry and diverse genetic makeup today.

In the United States, "Hispanic" is the name for a race made up of a biologically diverse assortment of Americans. It seems to be based on a common language and really nothing else. But languages are cultural, not biological. Interestingly, the "Hispanic" group is increasingly referred to as an "ethnicity." No doubt this is due to the clear absurdity of claiming that it is a category of people united by their biology. Yet people still think and speak of this group as if it were a biological race. What about "black" Americans and "white" Americans? Few societies in the world have such well-defined racial groups. Americans can barely make it out of their front door every day before someone wants them to identify their race on a form for one thing or another. But are "black" and "white" people really as different and biologically distant as it seems? Many "black" Americans have a significant amount of recent European ancestry. American hero Booker T. Washington wrote back in 1901: "How difficult it sometimes is to know where the black begins and the white ends."[43] It hasn't gotten any easier here in the twenty-first century; which brings up an interesting question. Why is President Barack Obama "black"? How does he get to be America's "first black president"? Race is supposed to be about biology, about genes, right? But Obama's father was "black" and his mother "white." So what happened to the mother's genetic contribution to her son? How can a woman give birth to a baby who ends up in a race that is different from hers? If race is about genetic *kinship* and *common* ancestry, then how can a mother be so distant from her child that the two aren't even in the same race? Is President Obama more closely related to other black people than he is to his white mother? How can "black" or "African American" be considered as the name of a distinct and valid racial group—based on biology—when one can belong to it with "white" ancestry of 50 percent or more? The answer, of course, is that race is not primarily about biology and ancestry. Races are cultural categories, created by social and historical circumstances.

Throughout most of US history it has been the norm for a person who had just one "black" grandparent and three "white" grandparents to be labeled "black." This is, of course, due to the belief that "white" blood is pure and that even small amounts of "nonwhite" blood result in contamination. This is not a relic of the past. Still today, many babies who are born to a white parent and a black parent are more likely to end up being identified as black. Clearly this shows that race in America is primarily a cultural game that is falsely advertised as biology.

THE SCIENTIFIC CASE FOR RACE

Adding to the challenge of raising public awareness about the reality of race is the fact that, still today in the twenty-first century, some scientists are hard at work trying to confirm that traditional race groups really do make scientific sense. Further complicating things, we may soon see more of an emphasis on race-based medicine. (This would be a big mistake, according to many scientists. The controversy is addressed in chapter 5.) There is also the problem of racism. Some people really are racist, and they are motivated to keep racial groups alive so that they can justify their worldview and keep the hate flowing. Even if it fails to prove anything conclusively, genetics may give support to race belief in the coming years. Jared Diamond warns, however, that new efforts to classify races based on genes are likely to lead to more unnecessary problems. "Racial classification didn't come from science but from the body's signals for differentiating attractive from unattractive sex partners, and for differentiating friend from foe. Such snap judgments didn't threaten our existence back when people were armed only with spears and surrounded by others who looked mostly like themselves. In the modern world, though, we are armed with guns and plutonium, and we live our lives surrounded by people who are much more varied in appearance. The last thing we need now is to continue codifying all those different appearances into an arbitrary system of racial classification."[44]

Our anatomically modern ancestors spent the last two hundred thousand years or so selecting mates across tribal lines, cultural lines, and any other lines they had half a chance to cross. For all our xenophobia and territoriality, we seem to make love not war (or both) in virtually every case of contact between groups. This consistent intergroup mating across thousands of generations around the world has led us to our present state, as a species with a wonderfully complex and tangled ancestral past. Whatever genetic research does or does not turn up, it can't possibly validate the popular concept of races. *Nothing* that comes out of the human genome can prove that traditional race belief is correct. Race belief, such as is practiced in the United States, is based on hopelessly illogical conclusions and inconsistent parameters. For example, no genetic discoveries can make biological sense out of the cultural rule that says one black parent or even one black grandparent makes a child black. No geneticist will ever emerge from her laboratory holding convincing evidence showing that the American view of race is valid but the Brazilian view of race is not. Race belief, as practiced by most people around the world today, is a collection of nonsense piled up on top of a foundation of fantasy and ignorance. No doubt there are many wonders to come from genetic research in the near future, but none of it can validate traditional race belief.

Anyone who is under the impression that recent genomic research has confirmed the existence of conventional races should be aware of this clear and straightforward statement from the Human Genome Project: "DNA studies do not indicate that separate classifiable subspecies (races) exist within modern humans. While different genes for physical traits such as skin and hair color can be identified between individuals, no consistent patterns of genes across the human genome exist to distinguish one race from another. There also is no genetic basis for divisions of human ethnicity. People who have lived in the same geographic region for many generations may have some alleles in common, but no allele will be found in all members of one population and in no members of any other."[45]

Science writer and author Olson keeps a close watch on new discoveries that have implications for human biology, origins, and diversity. "Recent genetic research has revealed that it's possible to get a decent sense of where many, and sometimes most, of a person's ancestors lived in the last few thousand years," Olson wrote to me in an e-mail message. "I personally don't see that as establishing the validity of biological races, though some have interpreted the results along those lines. Furthermore, research that I've done has shown that a person's genealogical ancestors rapidly expand, going back in time, to include everyone living in the past who has any ancestors living today."[46]

Another big problem with common race belief is that many people, probably most people, still are not aware of just how varied and overlapping our species is when it comes to observable physical features such as skin color, nose dimensions, and hair type. People still tend to oversimplify biological diversity, basing it all on a few relatively extreme examples in their minds. While traveling in Africa, Europe, South America, the Caribbean, and Asia, I paid close attention to the physical features that are traditionally used to assign people to racial categories. Being enculturated as American, I probably couldn't resist noticing these things even if my life depended on not doing so. Throughout my travels, I was surprised by how much diversity there was in the skin color, facial features, and hair type of the people I observed within each continent, within each country, and, yes, within traditional racial groups. Many white Europeans in Italy tend to look a little different from many white Europeans in France, for example. Many Africans in the Rift Valley tend to look different from many of the Africans in downtown Nairobi. People often refer to something called the "Asian race." But Asians in Nepal look different from many of the Asians I saw in Mumbai, India, or Bangkok, Thailand. Many Pacific islanders in Fiji look different from Pacific islanders in Papua New Guinea. Additionally, the more time I spent in a country, the more I noticed various patterns of physical features. I wasn't traveling with calipers to measure skulls or anything like that, so some of my observations may have been imagined or distorted by the random nature of the sample of people I encountered. But I'm confident that much of the diversity I saw reflected reality. This leads me to ask those who believe in biological race groups to explain how they can be content with five or six races. If you believe in races, then you can make the case that there are millions of races! This is one of the key problems with believing in racial groups the way most people do. Once you choose some trait or combination of traits to mark where one race ends and the next one begins, you have no logical reason to stop drawing boundaries around people. You can keep tweaking the extraction formula until you have thousands, perhaps millions, of races. Hey, maybe that's it. Races can finally make sense if we can just simply agree that every family in the world is its own distinct and separate race. But wait, that wouldn't work either, because at some point every family blends into another family—just as racial groups do. Back to the drawing board.

Another thing I noticed while traveling around the world and meeting people of every flavor is that many "racial looks" pop up in surprising places. There are people in Syria who look like many people I saw in Ecuador. And there are people in Fiji who look like many African Americans I know. Who knows what story their genes would tell, but based on their observable features these people were living, breathing violations of race belief. If race is supposed to be something that is obvious at a glance, as many race believers claim, then how can there be so many millions of contradictions in the flesh all around the world? For example, I could easily hand pick a few Fijian friends, dress them up in American-style clothes, and, so long as they didn't speak to reveal their accent, fool any American into believing that they were "black" or African Americans. I am confident that I could handpick 100 Caribbean people, take them to Fiji, and pass them off as indigenous Fijians in downtown Suva, as well.

The physical traits we rely on to create races are really there, of course, but what we make of them is our choice. If we want to see differences between white Italians and white French people, we can. If we want to see similarities between Fijians and African Americans, we can. What this means with regard to an objective analysis of the race question is that diversity is in the eye of the beholder. There is no natural breakup of humankind; there are no natural divisions.

Culture plays a big role in inspiring confidence in the validity of traditional racial groups. Language, dialect, dress, food, musical preferences, dancing styles, recreational interests, and so on all make a huge impression on us. These traits, although purely cultural, often fool us into thinking that they are determined by biology and are, therefore, racial. It is understandable that most people see biology lurking close behind all of these cultural traits because they have been wrapped up in the race package for centuries. One example is the "black" accent or dialect of many Americans who are members of the "black" or "African American" race. The way they speak has nothing to do with their DNA. It has to do with those whom they hear in infancy and childhood and after whom they pattern their speech. Undoubtedly, however, many people perceive a greater biological distance between "white" and "black" Americans because of dialect and other purely cultural traits. White Americans who are born and raised in Mississippi usually sound very different from white Americans who are born and raised in Boston. Obviously this is culture and not biology. There is no doubt that virtually any infant can be enculturated or raised up to speak any language, dress any way, or dance to any kind of music. But belief in race can mislead us. It suggests to us that all those differences in appearance, speech, and behavior are so deep and profound that they must be written into the DNA.

Yes, people can be very different. No one should dispute that. For example, walking the backstreets of Thailand and China and seeing the kinds of foods for sale surprised me. I knew diets were different around the world, of course. But I had no idea that deep-fried lizards, skewered scorpions, and freshly skinned snakes were common sidewalk food. Surprised as I was to see crunchy bats-on-a-stick, however, there is nothing written into the DNA about this. Had I grown up in rural China I might be nibbling on a tangy scorpion snack right now. It is important when thinking about race to always make the effort to mentally strip away all the things that are born of culture. Language, dress, diet, music, jewelry and other forms of body decoration, religious beliefs and practices, and so on. Once this is done, and we are naked, it's easy to see that we are more alike than different.

The United States and the rest of the world are likely to be much better off if the discrepancy between the way anthropologists view race and the way the public views race can be reduced. Today the typical layperson still thinks of race as it was thought of two or three centuries ago: of races as valid, well-defined categories of people objectively based on biology and thoroughly proven by scientific evidence. Most people in the world today are also likely to think that a wide range of behavioral,

personality, physical, and mental abilities are embedded in people biologically according to their race. And this position is by no means confined to
racists. Many well-meaning people who believe we should all hold hands
and live in peace are devout believers in biological races.

Meanwhile, many of today's scientists view race very differently.
When they talk about our biological diversity and the differences between
us, they speak about *clusters* or *clines* of traits bunching up in places along
an unbroken continuum of humanity. Many scientists would stress that
people are designated to races primarily by cultural tradition and incorrect
ideas about human diversity. They would also probably add that IQ,
morality, specific talents, and so on have never been proven to have a traditional racial component beyond the influence of culture—despite centuries of intense effort to prove otherwise.

There is an urgent need for public awareness to catch up to the basic
facts on which most scientists today are in agreement. It is not just wrong;
it is socially destructive for millions of people to continue believing that
membership in a racial group carries with it overwhelming biological
implications for an individual's ability and destiny within society. Yes, race
as a cultural force really does impact our lives. But we make the mistake of
thinking that it is nature, rather than culture, that empowers race. The
truth is that we give it undeserved power by misreading biological diversity and imagining that it places giant canyons between us. The scientists
are clear on this. "We are not saying that biological differences among
human groups do not exist, nor that racial differences are insignificant,"
explains Dr. Marks. "Differences among human groups do indeed exist, but
they do not sort the species into a small number of biologically fairly discrete groups."[47]

Chris Stringer, a leading paleoanthropologist, complains that it's difficult enough to get people to accept the basic scientific explanation of our
past, and so getting them to agree to the facts about our observable diversity is an even greater challenge. "Certainly for the public there is continuing confusion between biological and cultural meanings and uses of the
term [race]," says Stringer. "Human diversity is patterned geographically,
but not in a way that corresponds to any of the conventional 'racial' categories. Genetic diversity declines with distance of founder populations
from Africa. Hence any categorization that equates diversity in 'Africans'
with those outside of Africa is likely to be meaningless."[48]

"Meaningless" doesn't mean unpopular, however. Not only does the general public continue to believe in biological races, but Stringer adds that there is even a minority of anthropologists today who continue to use race. "For anthropologists, it is clear that there *is* meaningful genetic and physical differentiation in recent humans, and thus a minority still seem to prefer to use 'racial' categories as convenient—even if largely inappropriate—labels."[49]

No matter who does or does not believe in biological races, this much is certain: We are one species. Therefore, the potential and the value of any one member of our species should never be diminished or denied based on a concept that remains as contentious, elusive, and unproven as race.

THE TINY TRUTH WITHIN A GIGANTIC CLAIM

We have seen ways in which the popular concept of biological races falls short of matching the realities of our diversity. However, there remains the important question of whether or not races exist in any form at all. Is there at least some truth somewhere in the idea of races? The answer is yes. Races do exist—sort of. Obviously they exist as culturally created categories, but there also is some truth lurking deep within the biological theme behind conventional races. Anyone who declares that race is a total fabrication of the mind, with no connection to biological reality whatsoever, goes too far and gives aid and comfort to those who push full race belief. Physical traits that are observable by the layperson and genetic evidence that is detectable by science indicate that there is at least something to this thing called race. Popular race groups such as "white," "black," and "Asian" do correspond to geographical ancestry and kinship. The problem, of course, is that this grain of truth is microscopic and swamped by all the overlap, contradictions, and inconsistencies that readily appear when one takes a serious look at race. Maybe we need to come up with a new name that we can all use to refer to people without encouraging false beliefs and destructive attitudes. Or maybe that would be impossible. But no matter how desperately some may want to dump the concept of race because of its scientific shortcomings, or in the name of social harmony, we can't just declare it to be imaginary and then hope no one speaks of it ever again.

Anthropology, the branch of science that studies human culture and biology in the past and present, should know best of all whether or not races make sense. Currently, the discipline is not in unanimous agreement on the reality of races. While the majority of anthropologists reject biological races, there are some who do not. Some anthropologists continue to believe that races are valid and scientifically useful categories.[50] George W. Gill, a forensics anthropologist and University of Wyoming professor, favors the race-is-real side. What is most interesting about his position is that, although he argues that racial groups exist and should continue to be used by science, he freely admits that the race concept is complex, far from straightforward.

The following is an excerpt from an essay Dr. Gill wrote for the Web site that accompanied a PBS documentary, *Race: The Power of an Illusion:*

Where I stand today in the "great race debate"...is clearly more on the side of the reality of race than on the "race denial" side. Yet I do see why many other physical anthropologists are able to ignore or deny the race concept. Blood-factor analysis, for instance, shows many traits that cut across racial boundaries in a purely clinal fashion with very few if any "breaks" along racial boundaries. (A cline is a gradient of change, such as from people with a high frequency of blue eyes, as in Scandinavia, to people with a high frequency of brown eyes, as in Africa.) Morphological characteristics, however, like skin color, hair type, bone traits, eyes, and lips often, though not always, follow geographic patterns that coincide with climatic zones. This is not surprising since the selective forces of climate are probably the primary forces of nature that have shaped human races with regard not only to skin color and hair form but also the underlying bony structures of the nose, cheekbones, etc. (For example, more prominent noses humidify air better.) As far as we know, blood-factor frequencies are not shaped by these same climatic factors. So, serologists who work largely with blood factors will tend to see human variation as clinal and races as not a valid construct, while skeletal biologists, particularly forensic anthropologists, will see races as biologically real. The common person on the street who sees only a person's skin color, hair form, and face shape will also tend to see races as biologically real. They are not incorrect. Their perspective is just different from that of the serologist.

So, yes, I see truth on both sides of the race argument.[51]

LOOKING FOR THE RACE GENE

Some people thought that the mapping of the human genome and genetic research in general would prove to be the final nail in the coffin for race. That hasn't happened yet. In 2000, one of the big stories about the mapping of the human genome was the announcement that all humans are 99.9 percent alike. Race was on the ropes, fading fast, and appeared to be going down for the final count at last—at least within the scientific community if not the general public, which never doubted the validity of biological races in the first place. But then it was clarified that we may "only" be 99.0 percent alike,[52] that there is more genetic difference within our species than originally thought. Maybe this is new evidence of biological races. At least some of the people who bother to pay attention to the science news were probably left as confused as ever about the existence or nonexistence of this thing called race.

While anthropology has, for the most part, come down formally and decisively against the reality of biological races, science overall remains conflicted. Many psychologists, sociologists, and other researchers continue to use racial categories in ways that encourage people to continue believing in the validity of biological races. Some people go so far as to claim that new genetic research has established the existence of biological races once and for all. It hasn't. Meanwhile, others say it has shown that races do not exist. This is an arguable point as it is very difficult to prove the nonexistence of something. Someone can always say it exists in a slightly different way, for example. Many scientists who reject the race concept point out that 90 percent or more of human biological variation is found *within* the traditional races. Between the races there is variation of only 10 percent or less. This means that we can find people of the same race who are more different from each other than they are from someone of another race. "The remarkable feature of human evolution and history has been the very small degree of divergence between geographical populations as compared with the genetic variation among individuals," wrote anthropologist R. C. Lewontin and his coauthors in *Not in Our Genes* more than two decades ago.[53] This makes perfect sense in the context of prehistory and history. Travel and mating across cultural borders are not activities humans only recently picked up. We have been on the move and making babies at every step of the way for a long, long time. Our story is not one of different groups of people, isolated on their

own continents, with no connection to others. But while there may be relatively little genetic variation between the traditional races, that little bit of
difference between people with ancestries linked to different continents is
sufficient for some scientists to seize and use as justification for dividing
people into racial groups.[54]

A study led by geneticist Neil Risch gave hope to those who believe
genetic evidence will support the preservation of traditional racial groups.
Risch and his team studied DNA samples from 3,636 people who identified themselves as "African American," "white," "East Asian," or "Hispanic." The researchers then blindly matched the DNA samples to traditional racial groups. Only five of the samples did not match the racial
group the donors had selected for themselves.[55] This very low error rate
suggests that people know not only their cultural race but also their biological race. This is seen by some as powerful verification of the traditional
race concept. I would point out, however, that the world is far more complex than 3,636 people who identify themselves as "African American,"
"white," "East Asian," and "Hispanic."

Armand Marie Leroi, an evolutionary biologist, supports the concept
of race because he believes that some genetic traits—when analyzed as
correlations—reveal races. "Race is merely a shorthand that enables us to
speak sensibly, though with no great precision, about genetic rather than
cultural or political differences," he wrote in a 2005 *New York Times* opinion
essay. "But it is a shorthand that seems to be needed. One of the more
painful spectacles of modern science is that of human geneticists piously
disavowing the existence of races even as they investigate the genetic relationships between 'ethnic groups.'... [T]he recognition that races are real
should have several benefits. To begin with, it would remove the disjunction
in which the government and public alike defiantly embrace categories that
many, perhaps most, scholars and scientists say do not exist."[56]

Such words draw strong reactions from some scientists. They are concerned by what they see as a cavalier attitude toward the matching of
genetic data with cultural categories called races. In the words of biologist
Ruth Hubbard: "It is beyond comprehension, in this century which has witnessed holocausts of ethnic, racial, and religious extermination in many
parts of our planet, perpetrated by peoples of widely different cultural and
political affiliations and beliefs, that educated persons—scholars and pop-
ularizers alike—can come forward to argue, as though in complete inno-

cence and ignorance of our recent history, that nothing could be more interesting and worthwhile than to sort out the 'racial' or 'ethnic' components of our thoroughly mongrelized species so as to ascertain the root identity of each and everyone of us. And where to look for that identity if not in our genes?"[57]

Looking at what both the pro-race and no-race anthropologists agree on may be the best way to find something the public can confidently believe to be true about race. There is wide agreement among anthropologists today about the following: We became anatomically modern people in Africa about two hundred thousand years ago. At some time between one hundred thousand and fifty thousand years ago, we spread throughout Africa and beyond, eventually reaching every continent except Antarctica. As common sense would suggest, groups of people who lived in the same region would probably be more closely related to each other than they are to groups of people who lived at that time in very distant and different environments. These are the key facts on which proponents of racial groups base their case. They say some populations reflect a shared ancestry. What may be surprising about this is that anthropologists who reject race say the same thing. It turns out that the two warring sides of the race issue are often closer to one another than it seems at first glance. This is a good sign for those who hope to understand race for what it really is. Yes, there are still battles raging over the use of the term "race" and over claims that biological race is relevant to intelligence, for example. At least, however, there is the potential for total agreement on the most basic facts about our biological diversity. This is a start.

Some people suspect that scientists who argue in favor of the existence of traditional racial groups are racists who are trying to support their social views by manipulating science. The flip side of that is that those who say biological races don't exist are sometimes accused of being overzealous liberals who are trying to support their social views by manipulating science. Of course while the scientists bicker, the pubic continues to work from a race template that was built on folktales, false assumptions, fears, ignorance, and eighteenth-century pseudoscience. Things might be much better if the public could at least become aware of the points that race believers and race rejecters agree on. For example, today there is no dispute among credible scientists that humankind is one species; that traditional racial groups do not represent pure races; and that there is great diversity within traditional racial groups. The real conflict comes when questions arise about the alleged borders between populations; the number of races; whether or not races are useful for the study and understanding of our biological diversity; and whether or not biological race plays a role in inherited intelligence, innate athletic ability, or genetic health. All these points will be addressed in detail in later chapters. For now, we are concerned with whether or not one is justified in saying that races exist at all.

First we have to confront the question of definition. How one defines race is key to making any progress. Failure to define what everyone is talking about inevitably leads to confusion. It is impossible to have a sensible conversation or debate about the existence or nonexistence of races if there is no agreed-upon definition of race in the first place. For example, one person may think of races as vague clusters of people who are related to a degree but not necessarily with great significance. Meanwhile, another person may refer to a popular definition that says races are specific, well-defined biological categories of people with profound implications for intelligence and moral behavior. These two people would be working from very different points of view and the quality of their exchange would suffer as a result. The first thing they need to do is debate the definition of race before they debate the reality of races. As polarized as the issue may seem at times, there is significant common ground between the two sides of the race debate, at least within the academic community.

One of the most important messages that the public could hear about race today is that virtually no modern scientists—whether they are liberal, conservative, pro-race, anti-race, racist, or otherwise—believe in racial categories as strictly defined units into which we can objectively and consistently place all of the world's people. Even devout race believers admit that the walls between their races are porous. For example, anthropologist Vincent Sarich is well known in the science subculture for his position that biological races are real and have real-world consequences for intelligence, sports ability, and so on. He has irritated and enraged many colleagues and students with these controversial claims.[58] But, putting those conflicts aside for the moment, it is important to know that Sarich does not define race in the same way the guys down at your local sports bar probably do. He doesn't think of race in the same way members of the neighborhood bridge club do, or most journalists do, or most people in positions of power and influence within governments do. Dr. Sarich may be a believer in races, but when it comes to the definition of race, he differs significantly from the public. He is almost as misaligned with the popular notion of race as the staunch no-race anthropologists are. He knows, for example, that when it comes to race, humankind is at most broken up into "fuzzy sets."[59]

I met Dr. Sarich back in the late 1980s when he gave a guest lecture to my physical anthropology class. Two things stand out in my memory of him: (1) he is extremely tall, and (2) he speaks freely, without much concern for what might be called political correctness. I also attended a creationism versus evolution debate between Dr. Sarich and Duane Gish of the Institute for Creationism Research. As I recall, Dr. Sarich destroyed Gish with logic and evidence but still lost the crowd, at least in part because he didn't seem much concerned with winning them over. Sarich rose to fame as a young graduate student in the 1960s, when he teamed up with the late Allan Wilson to use genetic material to recalibrate our evolutionary timeline. Their groundbreaking work in creating a molecular clock showed that human ancestors split from other African apes as recently as five million years ago, rather than twenty million years ago, as was commonly believed at that time. Today, because of his views on race, Sarich is vilified by many in the academic world as a right-wing racist who is groping for science to validate his evil views. Maybe that's true, maybe it's not. Who can ever really know the answer to that, other than Vincent Sarich? For our purposes here, what is going on in his head or in the head of any other scientist is not the priority. This book is concerned with ideas and evidence concerning race, not with the possible secret motivations of individuals. So, how does Dr. Sarich define race? In their book, *Race: The Reality of Human Differences*, he and co-author Frank Miele write: "there is a substantial amount of agreement in the field on a working definition of the term 'race.' ... Races are populations, or groups of populations, within a species, that are separated geographically from other such populations or groups of populations and distinguishable from them on the basis of heritable features. ... Everyone can agree that at one level we are all members of a single species—*Homo sapiens*—and that at another each of us is a unique individual. Races then exist within that range to the extent that we can look at individuals and place them, with some appreciable degree of success, into the area from which they or their recent ancestors derive."[60]

This dry, rather noncontroversial definition of race comes from a man who is public enemy number one in the eyes of many no-race anthropologists. Sarich may deserve much of the heat he gets for his support of ideas about race and intelligence but certainly not for this bland definition of race. It is sensible and defensible, to a point at least. It also sounds familiar. Recall the words of Jonathan Marks, the prominent no-race biological anthropologist quoted earlier in this chapter: "We are not saying that biological differences among human groups do not exist, nor that racial differences are insignificant. Differences among human groups do indeed exist, but they do not sort the species into a small number of biologically fairly discrete groups."[61]

Notice the similarities here. These men are supposed to represent two polar opposites when it comes to the concept of race—and they do on many key issues—but when it comes to the most basic descriptions of human diversity they seem to be rather close to one another. Sarich and Marks agree that races (populations, groups, clines, or clusters, if you prefer) exist at least in the form of some genetic relationships between people with ancestors linked to the same geographical area. For another example, consider Arthur Jensen, the psychologist who has promoted the idea that mental ability is determined in large part by race. His ideas have been highly controversial, to say the least. What does he think race is? "I am in complete agreement with one important point in the AAA [American Anthropological Association] statement [on race], and I don't know anyone who is up on this subject who would disagree. Races are not biologically clear-cut categories or distinct groups.... races are 'fuzzy' groups with clines, or blends, at their blurry boundaries."[62]

This is another case where there is some level of agreement on the basic facts. Again, sadly, the odd person out is the typical member of the public who has heard nothing of the idea that races are not natural, well-defined biological groups of people. The layperson, here in the early twenty-first century, is likely still to believe that races are naturally occurring categories people are born into in a manner ordained by a god or nature, and that much of their life's destiny is largely sealed as a result. Racial identity is still a big deal to most people.[63] Many are also likely to believe that science has somehow verified races as being real beyond a doubt. Members of the public may or may not be racists, but they are likely to believe that races exist in a way that modern science has universally rejected.

The gap regarding the definition of race that exists between the scientific community and the public at large is critical to the race problem in

America and throughout the world. If everyday people on the sidewalks knew a few basic facts about our biological diversity, as anthropologists even as diverse as Sarich and Marks lay it out (that is, we are one species, we are all unique individuals, there are no natural fixed walls between groups of humans), then we might begin to see significant movement away from races as these tragic psychological prisons that inspire and fuel so much misinformation, mistrust, discrimination, violence, and wasted potential.

Jefferson M. Fish, professor emeritus of psychology at St. John's University, is a rare psychologist, having made a serious investigation into the concept of race and emerged a nonbeliever. He is the editor of an important and influential book, *Race and Intelligence: Separating Science from Myth*. Fish was friendly and patient during our long discussion, but several times I sensed his frustration with fellow psychologists and the world in general for not "getting it" when it comes to the myth of biological races.

"The idea of races is that the human species started to split up early on but that is just not what the pattern has been," Fish said:

> The pattern has been not of a branching tree but like a tangled lattice where groups have broken apart and come back together with others. What you have is a pattern of gradual variation around the world rather than distinct separation. That's one half of the problem; the other half is that you can look from culture to culture and see that race means different things to different people.
>
> When you tell people that there aren't races, they seem to think that you're blind. "Can't you see that people don't all look alike?" Of course people vary in what they look like. It's just that the cultural notion of who goes together in a category called a race doesn't match up with the biological facts. People can change their race just by getting on a plane and flying to a different culture. What changes is not what they look like, not their genes, not their ancestry, but rather their cultural category changes. [For example,] the American system of race is based on ancestry and the cultural word that is used for that is "blood." When someone in the United States tells you that they are black, you have no idea what they look like. All you would know is that they have an ancestor who is classified as black. In contrast, in Brazil, they have what they call "tipo" [type]. That's a constellation of physical features. Brazil has a lot of variation, especially compared to the United States. There was a study done in which Brazilians were asked, "What color are you?" and they got 134

different answers. One of the things Brazilians say about Americans is that we are racists because we call people "black" who aren't "black." People like Colin Powell or Malcolm X, for example, would not be considered "black" in Brazil.[64]

Fish added that no one culture is correct and all the others wrong about racial classifications. "These are just different kinds of classifications which grew up for historical, cultural, and linguistic reasons, not for biological reasons," he explained.[65]

Fish hopes that more basic education about what races really are will improve the situation, especially for younger children. However, he does not believe that this will necessarily stop people from separating and hating. "Unfortunately, there are so many things we use to divide [up] ourselves over that it seems like it will never end. But that is no reason for us not to try and do a better job with the concept of race."[66]

Trying to get a firm grip on the concept of race can be frustrating for the layperson who makes the effort to investigate it. Sometimes it's like trying to squeeze water and hold it in your hand. One day we read about an anthropologist who says biological races are not real, and then, the next day, we read about a geneticist who says they are real. Strangely, both scientists could be correct, technically, because they may be working from different definitions of race. One could mean that race as a system of unique, well-defined biological categories based on consistent traits is not real, while the other means that races do exist in the form of generalized populations or clines that share similar genetic ancestries. Confusing, yes, but they may not necessarily be contradictory positions. One could argue that such confusion is a very good reason to abandon all use of the world "race" in scientific work and instead find a nice technical term to use, one that doesn't encourage misguided race belief in the public.

Yes, there is something to this thing called race. We are not imagining things if we notice that a Swede, an Ethiopian, and a Chinese person standing side by side have physical differences that are noticeable and can be attributed to their respective ancestries. But in acknowledging the realities of those differences, we need to understand that this goes only so far. The world is made up of many more people than three individual samples could ever hope to represent. The idea of races may seem simple, but human diversity is far too complex to reduce to a three-person lineup.

There are no pure races. There are no ideal examples within any race. And, most important, there are no biological barriers or borders between conventional race groups. Finally, as humans, the Swede, Ethiopian, and Chinese person would have far more in common than a devout race believer might be inclined to notice.

There has always been a gradual blending of one population, or race, into the next. Yes, scientists may be able to use genetic information to identify groups within the human population that share some degree of common ancestry, but this can lead down many different paths, and what you find may depend on what you're looking for. Scientists may even be able to use specific genetic information in order to assign an individual to an ancestral group or race.[67] While this may show that there is something that vaguely hints at race, it does not confirm the traditional notion of race. For example, as genetic research progresses, it is likely that scientists, using genetic information only, will be able to place an individual not just in a large group that corresponds with some vast traditional racial group but also in a very small, specific group of people. If this happens, and it surely will, those who defend race may find themselves in the position of having to defend not several races but several thousand races. We may end up with so many races that the term loses all meaning, even for those who have believed in it the most. Perhaps the eventual result of this path will be a final arrival at a place where we are all seen as individuals. Imagine that: every man, woman, and child as a race unto himself or herself.

REJECTING RACE FOR THE RIGHT REASONS

We can't wish away race because we don't like racism. If we bury race in a flurry of good-hearted, well-intentioned social engineering, it will claw its way out of the soil and come back to haunt us. If we hope to fix the race problems in society, then we first need to achieve a broad public understanding of race that includes a scientifically valid definition. No one should be uptight about acknowledging the facts of our species and its diversity. There is no contradiction in accepting that popular perceptions of race are wrong while also accepting that there are biological relationships between people that vary in degree depending on how close or far apart their ancestors have been over the last 50,000 years. The crucial challenge is to understand as much as possible about biological diversity while making sure to challenge the unfounded ideas and exaggerated claims that justify racism. We also need to be mature as a species and recognize how shortsighted it is to obsess over a small pool of ancestors who supposedly were on one continent twenty thousand years ago, rather than focusing on our ancestors who were all on the same continent one hundred thousand years ago.

Understanding and accepting who we are is a worthy goal. It is my belief that we can replace traditional race belief as it exists in the minds of most people today with scientifically accurate facts about human diversity and be better off for it. Ultimately this is simply about teaching science. Our destiny must be where the evidence leads. If tomorrow someone discovers overwhelming evidence that establishes the reality of discrete and logical racial categories, then my next book probably will be about the confirmed existence of biological races. We don't have to ignore our differences. We don't even have to abandon our socially constructed races if we don't want to. We just have to think of them in a new way, an honest and realistic way. We really can have it both ways. We can jettison the traditional concept of race while holding on to whatever aspects of cultural and biological diversity we enjoy.

Hopefully scientists will continue, doing more rather than less, to study human diversity and our evolutionary past. Yes, the efforts of many anthropologists in the past were terribly misguided, sometimes to the point of inspiring and aiding pure evil, but this should not condemn forever the worthy goal of trying to find out as much about ourselves as we can. Some people may feel that research related to diversity or race should be discouraged because it's linked to traditional and dangerous ideas about race and is therefore too risky. However, the problems with race that we have endured historically have been the result of ignorance. Government leaders and the public they govern have always known too little about the reality of our biological diversity rather than too much. Continued ignorance is likely to ensure racism in the future. My guess is that one day we will learn enough about our biological past and present to conclude that racial groups can be said to exist, but in such a limited and meaningless form that they have no practical use for society or science. Scientists will probably refine ways to extract precise biological identities from people, but this will lead to more and more "races" being named until this reaches a point of absurdity. Perhaps soon after the day arrives when everyone can afford to have his or her personal genome sequenced, the concept of biological races will drown in a sea of individualized humanity. I suspect that races are doomed to become meaningless in day-to-day life. Racialized medicine may become very popular, for example, but if it does, it will probably be promptly replaced by individualized medicine, given the rapid advancement of genomic research. We do not need to fear race, run away from it, or pretend we don't see biological differences. We need to understand what our diversity is and, most important, what it is not.

Whatever kernel of truth there may be within the concept of race, Dr. Marks is unwilling to concede much if anything to those who seem determined to keep the traditional concept of race alive in science and among members of the public. "Races, as natural divisions of the human species, are thus rather like angels," says Marks. "Many people believe in them, devoutly. They can even tell you what properties they have. But the closer you try to examine them to discover their real nature, the more elusive they become. And ironically, the people who claim to be most familiar with them are the ones to be the most suspicious of."[68]

Even after reading countless books and papers about our biological diversity and evolutionary history; after sitting attentively through many anthropology lectures; and after traveling the world in search of race, I still struggle to concisely define it. It is one of the greatest contradictions of all time. Race means so much, yet so little. It is so real and powerful, yet it's not really there.

What is it?

If I were pressed to condense my thoughts about race, I would probably come up with something like this: Based on all we know about our past and our diversity so far, race seems too small a word to contain my story, much less the story of my entire species. Some people may call me "white" but everyone calls me "human." That means my story stretches all the way back to a small band of the first modern people in Africa. All the points of geography and time that lie between now and then matter, but none of them are so important as to be able to define me in total.

There is too much to fit inside this little word, "race." All of the inconsistencies, twisted truths, and cultural contaminations invalidate it. The weight of my past and your past squashes it to nothing. Race does not exist in the way most people believe it does. But that doesn't mean it's not real.

Of course race is real; we made it real by believing in it. But that's culture, not nature. So to know me or anyone else, don't look for race. Look for my story, the real story.

Does accepting that biological races are not real mean we must lose something good along with the bad? Does turning our back on race belief and racism mean losing the richness of diversity? No, biological races are not real, and that is never so clear as when they go away. I let go of race belief long ago, but I never let go of our beautiful biological diversity. I also never felt the need to deny the reality of our cultural variation or the importance of cultural groups for many people. In fact, moving past race allowed me to embrace the world's diversity more than I could ever have done as a race believer. I have seen the world, far from the beaten path of tourists, and, as a result, I have a pretty good idea of how different people can be. However, even as I marveled at our differences in both culture and biology, I could never escape the reality before my eyes. We are so different, yet so alike. There is nothing contradictory in that statement. It is who we are, a collection of differences created out of sameness.

NOTES

1. Luigi Luca Cavalli-Sforz and Francesco Cavalli-Sforz, *The Great Human Diasporas: The History of Diversity and Evolution* (New York: Helix Books, 1995), p. 237.

2. Joseph L. Graves, *The Emperor's New Clothes: Biological Theories of Race at the Millennium* (New Brunswick, NJ: Rutgers University Press, 2001), p. 5.

3. "The Difference between Us," episode 1 of *Race—The Power of an Illusion*, PBS, 2003, www.newsreel.org/transcripts/race1.htm (accessed September 14, 2009).

4. Jared Diamond, "Race without Color," *Discover*, November 1994, http://discovermagazine.com/1994/nov/racewithoutcolor444 (accessed September 14, 2009).

5. Charles N. Rotimi, "Are Medical and Nonmedical Uses of Large-Scale Genomic Markers Conflating Genetics and 'Race'?" *Nature Genetics* 36 (October 26, 2004): S43–S47, www.nature.com/ng/journal/v36/n11s/full/ng1439.html (accessed September 14, 2009).

6. James Shreeve, "Terms of Estrangement," *Discover*, November 1994, http://discovermagazine.com/ (accessed September 14, 2009).

7. Leonard Lieberman, Blaine W. Stevenson, and Larry T. Reynolds, "Race

and Anthropology: A Core Concept without Consensus," *Anthropology and Education Quarterly* 20, no. 2 (June 1989): 67–73.

8. Heather J. H. Edgar and Keith L. Hunley, "Race Reconciled? How Biological Anthropologists View Human Variation," *American Journal of Physical Anthropology* 139, no. 1 (May 2009): 2.

9. Ibid., p. 4.

10. American Anthropological Association, "American Anthropological Association Statement on 'Race,'" May 17, 1998, www.aaanet.org/stmts/race pp.htm (accessed September 14, 2009).

11. "Genetically Speaking, Race Doesn't Exist in Humans," Eurekalert.org, www.eurekalert.org/pub_releases/1998-10/WUiS-GSRD-071098.php (accessed March 30, 2009).

12. Association of Physical Anthropologists, "AAPA Statement on Biological Aspects of Race," *American Journal of Physical Anthropology* 101 (1996): 569–70, http://physanth.org/position-statements/biological-aspects-of-race (accessed September 14, 2009).

13. Curtis W. Wienker, interview by the author, January 23, 2009.

14. Ibid.

15. Ibid.

16. Ibid.

17. Milford H. Wolpoff, "How Neandertals Inform Human Variation" (unpublished paper).

18. Nick Wynne, interview by the author, April 6, 2009.

19. Yehudi Webster, *The Racialization of America* (New York: St. Martin's Press, 1992), p. 91.

20. Steve Olson, interview by the author, April 22, 2009.

21. American Sociological Association, "The Importance of Collecting Data and Doing Social Scientific Research on Race," 2003, http://www2.asanet.org/media/asa_race_statement.pdf (accessed September 14, 2009).

22. Brandon Keim, "Poverty Goes Straight to the Brain," *Wired Science*, March 29, 2009, http://blog.wired.com/wiredscience/2009/03/poordevelopment.html (accessed September 14, 2009).

23. Curtis W. Wienker, "The Anthropological Perspective on Race: An Historical Overview," *American Association of Behavioral and Social Sciences Journal* 4 (Fall 2001), http://aabss.org/journal2001/wienker2001.jmm.html (accessed September 14, 2009).

24. American Anthropological Association, "Early Classification of Nature," in *Race: Are We So Different?* a project of the American Anthropological Association, 2007, http://understandingrace.org/history/science/early_class.html (accessed September 14, 2009).

25. Audrey and Brian D. Smedley, "Race as Biology Is Fiction, Racism as a Social Problem Is Real," *American Psychologist* 60, no. 1 (January 2005): 19.

26. Wienker, "The Anthropological Perspective on Race."

27. Ashley Montagu, *Man's Most Dangerous Myth: The Fallacy of Race* (Walnut Creek, CA; AltaMira, 1997), p. 48.

28. Ibid., p. 176.

29. Wienker, "The Anthropological Perspective on Race."

30. Ibid.

31. Steve Olson, *Mapping Human History* (Boston: Houghton Mifflin, 2002), p. 179.

32. Jonathan Marks, *What It Means to Be 98% Chimpanzee* (Berkeley: University of California Press, 2002), p. 94.

33. Jonathan Marks, interview by the author, January 25, 2009.

34. Vincent Sarich, "In Defense of the Bell Curve," *Skeptic* 3, no. 3 (1995): 85.

35. Shreeve, "Terms of Estrangement."

36. Graves, *The Emperor's New Clothes.*

37. Joseph L. Graves, *The Race Myth: Why We Pretend Race Exists in America* (New York: Dutton, 2004), pp. xiv–xv.

38. Jonathan Marks and Milford H. Wolpoff, personal communications with the author.

39. Stephen Molnar, *Human Variation* (Upper Saddle River, NJ: Prentice Hall, 1998), p. 1.

40. Ibid., p. 22.

41. Olson, *Mapping Human History*, pp. 33–34.

42. Diamond, "Race without Color."

43. Booker T. Washington, *Up from Slavery*, 1901, http://www.bartleby.com/1004/6.html (accessed September 14, 2009).

44. Diamond, "Race without Color."

45. Human Genome Project, "Minorities, Race, and Genomics," www.ornl.gov/sci/techresources/Human_Genome/elsi/minorities.shtml (accessed April 25, 2009).

46. Steve Olson, interview by the author, March 4, 2009.

47. Portuguese Marks, *Human Biodiversity: Genes, Race, and History* (New York: Aldine de Gruyter, 1995), p. 112.

48. Chris Stringer, interview by the author, March 4, 2009.

49. Ibid.

50. George W. Gill, "Does Race Exist? A Proponent's Perspective," *Nova Online*, PBS, November 2000, www.pbs.org/wgbh/nova/first/gill.html (accessed September 14, 2009).

51. Ibid.

52. Malcolm Ritter, "Study: Humans' DNA Not Quite So Similar," *Red Orbit*, September 4, 2007, www.redorbit.com/news/science/1054012/study_humans _dna_not_quite _so_similar/index.html (accessed September 14, 2009).

53. R. C. Lewontin, Steven Rose, and Leon J. Kamin, *Not in Our Genes: Biology, Ideology, and Human Nature* (New York: Pantheon Books, 1984), p. 127.

54. Michael J. Bamshad and Steve E. Olson, "Does Race Exist?" *Scientific American*, November 10, 2003, http://www.scientificamerican.com/article.cfm?id =does-race-exist (accessed September 14, 2009).

55. Stanford University Medical Center, "Racial Groupings Match Genetic Profiles, Stanford Study Finds," *ScienceDaily*, January 31, 2005, http://www .sciencedaily.com/releases/2005/01/050128221025.htm (accessed September 14, 2009).

56. Armand Marie Leroi, "A Family Tree in Every Gene," March 14, 2005, www.nytimes.com/2005/03/14/opinion/14leroi.html?_r=1&oref=slogin (accessed September 14, 2009).

57. Ruth Hubbard, "Race and Genes," in *Is Race "Real"?* a Web forum organized by the Social Science Research Council, June 7, 2006, http://raceand genomics.ssrc.org/Hubbard/ (accessed September 14, 2009).

58. *New York Times*, "Campus Life, Berkeley: Campus Is Split over Statements by a Professor," December 23, 1990, http://www.nytimes.com/1990/12/23/style/ campus-life-berkeley-campus-is-split-over-statements-by-a-professor.html (accessed September 14, 2009).

59. Vincent Sarich and Frank Miele, *Race: The Reality of Human Differences* (Boulder, CO: Westview, 2004), p. 209.

60. Ibid., pp. 207–208.

61. Marks, *Human Biodiversity*, p. 112.

62. Frank Miele, *Intelligence, Race, and Genetics: Conversations with Arthur Jensen* (Boulder, CO: Westview), p. 118.

63. Associated Press, "Survey Indicates White People Have Sense of Racial Identity," September 11, 2006, http://kaal.dayport.com/article/view/104450/ (accessed September 14, 2009).

64. Jefferson M. Fish, interview by the author, March 30, 2009.

65. Ibid.

66. Ibid.

67. Bamshad and Olson, "Does Race Exist?"

68. Jonathan Marks, "The Realities of Races," in *Is Race "Real"?* a Web forum organized by the Social Science Research Council, June 7, 2006, http://race andgenomics.ssrc.org/Marks/ (accessed September 14, 2009).

Chapter 2

WE ARE ALL AFRICANS

It is a common practice of humans to identify their prejudices with the laws of nature.

—Ashley Montagu[1]

The influence of the human evolutionary past and the results of ignorance of that past are nowhere more obvious than in the area of race, and few kinds of ignorance have had more pernicious effects in society. Racial prejudices stand, among other things, as a major barrier to the cooperation that will be required if the human predicament is to be resolved. Neither global nor regional and local environmental problems are likely to be solved if different groups are battling one another on the basis of imagined differences in human "quality."

—Paul Ehrlich[2]

Groups vary in appearance because their ancestors had different biological histories. But just how different were those histories? What if the distinctive appearances of human groups are a historical accident, a biological joke, no more substantial than masks at a costume party?

—Steve Olson[3]

We are all cousins—everyone on Earth.

—Carl Sagan[4]

If we want to have an understanding of the origins and realities of our present biological diversity, then it's a good idea to know something about prehistory. If, as many people believe, races are biological categories that were imposed upon us by nature, then we have to look at how this may or may not have happened. To do this we need to have at least a passing familiarity with the human story as it is described by the best current scientific evidence.

Where are we from? How did we get here? How did people come to inhabit so many places around the world? Why do people with some or most ancestors in one continent tend to look different from people who have ancestors in another continent? Do all people share a common origin? Does the story of prehistoric humans strengthen or undermine the concept of races as biological categories?

The first thing to address is time. The human story won't make much sense without a basic understanding of how long it is and where it fits in with everything else. It is important to at least have a basic timeline of key events and time spans in one's mind. This can be a challenge, no doubt. We normally think in days, weeks, and years. But to understand something about the universe, life on Earth, and humankind we have to be able to think in terms of hundreds of thousands, millions, and billions of years. This problem of time is important because I often sense or directly hear from people that our current "racial differences" have "always" existed. This is understandable, perhaps, because if people do not have an understanding of a basic timeline, superficial differences can gain unjustified importance. People with no idea of how, why, or when differences in skin color, hair, and facial features came about are more likely to believe falsehoods about their significance. Once we know the human story and

Important Time References:

- The universe is 13.7 billion years old;
- The Earth is 4.5 billion years old;
- Life on Earth began at least 3.4 billion years ago;
- The first vertebrates (animals with backbones) appeared in the ocean some 500 million years ago;
- The first land animals evolved about 400 million years ago;
- *Australopithecus*, a human ancestor, lived in Africa 2 to 4 million years ago;
- Based on current evidence, *Homo habilis*, *Homo ergaster*, and *Homo erectus* evolved in Africa some 2 million years ago;
- *Homo erectus* spread into Europe and Asia but is believed by most paleoanthropologists to have become extinct;
- Based on current evidence, anatomically modern humans (people like us) evolved in Africa approximately 200,000 years ago;
- Some of these anatomically modern humans migrated out of Africa beginning 100,000 to 50,000 years ago;
- Anthropologists believe that physical differences in skin color, facial features, and hair type evolved only over the last several thousand years in response to different environments;
- The first civilizations arose in North Africa and the Middle East approximately 5,000 years ago;
- The Atlantic slave trade thrived from the sixteenth century to the nineteenth century. Africans, mostly western and central, were taken to Europe, North America, South America, and the Caribbean to work under brutal conditions. Millions of lives were lost on slave ships during the infamous "Middle Passage." Many more millions of Africans and their descendants went on to endure brutal working and living conditions. The concept of race became a useful explanation and excuse for this massive international trade and the believed inferiority of "black people";
- During the eighteenth, nineteenth, and twentieth centuries, anthropologists and other scientists sought proof of white racial supremacy and the inferiority of other races, using, for example, skull measurements and intelligence tests;
- Anthropologist Ashley Montagu published his powerful book, *Man's Most Dangerous Myth: The Fallacy of Race*, in 1941. In it he detailed many factual and logical errors concerning belief in biological races;
- In 1996, the American Association of Physical Anthropologists published a statement on race. It included the following: "Pure races, in the sense of genetically homogenous populations, do not exist in the human species today, " and "Racist political doctrines find no foundation in scientific knowledge concerning modern or past human populations";
- In 1998, the American Anthropological Association released its own statement on race, which declared that biological races are not real; that races are cultural creations;
- In the year 2000, a working draft of the human genome was completed. To date, nothing has been found that confirms the traditional belief in biological races.

understand how we have changed over time, it is much easier for us to see that something like straight or curly hair is not a sensible reason for splitting up our species. (Refer to the timeline on the previous page.)

As with everything in science, these dates and time spans are open to revision should contradictory evidence come along. While it may not be vital to keep all of these numbers in your head, I do think it is important to retain an idea of the sequence and general time spans. Try to achieve an overall feel for the immense age of the universe, the age of the Earth, and the brief flash of existence that we humans represent. Know that there was no Earth during most of the existence of the universe. For the overwhelming majority of the ages in which there has been life on Earth, microbes were the only game in town. All of human civilization fits into a shockingly small sliver of time compared to what has come before us. The time since the Industrial Revolution seems nonexistent in comparison with 13.7 billion years, the age of the universe. If nothing else, all of this should encourage us to be humble.

EVERYTHING IN A YEAR

The late Carl Sagan's "Cosmic Calendar" is an excellent way to visualize huge time spans and put everything in perspective. In his book, *Dragons of Eden*, Sagan compressed everything—from the Big Bang to the present—into something we are all familiar with, a single 12-month year. This means the Big Bang occurs at the start of New Year's Day and the final moment of New Year's Eve twelve months later is the present day. By placing significant events on dates relative to their position in real time, we can easily see how very old some things are (the Milky Way Galaxy) and how very young other things are. Humans, for example.

Discovery Education and the American Museum of Natural History have an online version of the Cosmic Calendar, titled "The Universe in One Year," that is well worth exploring.[5]

The following time line is based on "The Universe in One Year." Remember, this is where key events would fit if the entire history of the universe—everything—were squeezed into a single year. Each month represents more than a billion years. Be warned, you will never look at a clock the same way again.

THE UNIVERSE IN ONE YEAR[6]

January 1 (New Year's Day): The Big Bang, the formation of the universe;

March: Our galaxy, the Milky Way, comes into being;

August: The Sun and planets of our solar system come into being;

September: Oldest known life on Earth;

December 15: Cambrian Explosion, the rise of many new life forms on Earth;

December 17: First vertebrates (animals with backbones);

December 18: Dawn of land plants;

December 21: Insects flourish;

December 24: First dinosaurs;

December 31 (New Year's Eve, the last day of the year):

9:24 p.m.—First human ancestors walk upright;

10:48 p.m.—*Homo erectus* reigns as the smartest creature on Earth;

11:54 p.m.—Anatomically modern humans appear;

11:59:45 p.m. (fifteen seconds before the present)—Writing invented;

11:59:50 p.m. (ten seconds before the present)—Egyptian pyramids built;

11:59:59 p.m. (one second before the present)—Voyage of Christopher Columbus.

Mind boggling, isn't it? Everything that has happened since Christopher Columbus crossed the Atlantic five centuries ago would be crammed into *a single second* in the perspective of this "Universe in a Year." While we can say we have made a big splash here on Earth, that splash has been incredibly short lived so far. Our galaxy was formed in March and the solar system came together in August. The first life on Earth didn't even show up until September, the ninth month. Dinosaurs don't make the scene until December 25. We don't make an appearance until a couple of hours before the end of the year! As for the last hundred years or so, all the events, discoveries, inventions, and wars that we think so much of took place in the last few tenths of a second.

Now it should be easy to see that, in comparison with the age of the universe, we live on a young planet. In comparison with the first arrival of life, humankind is a very young species. In comparison with our entire prehistory, human migrations around the world are very recent. Understanding time plays an important role in the understanding of race, because so much of the race idea today is rooted in the false perception that human biological diversity is something permanent and unchangeable. Many people hold the view that diversity or race is "just the way it always has been and always will be." There is no need to question race, some feel, because it is an unassailable feature of reality that cannot be changed or explained away. The cosmological perspective—the real story—shows something very different, however. A mere couple of hundred thousand years ago we were no more than a handful of souls struggling for survival in Africa. At that time, all modern people, all of our ancestors most likely had the same skin color, the same noses, the same hair type. (Isn't it odd that such things are deemed so important that we have to point this out?) Most of the observable traits that race belief depend on so heavily (skin color, hair texture, body shape, facial features, and so on) appeared recently—very recently. On the Cosmic Calendar these differences would have arisen within the last "four minutes" or so of the year. Most of these differences emerged after we expanded into diverse environments. This crucial migration event, taking place about fifty thousand years ago, led to much of the observable diversity that today's racialized worldview is based upon. But when those first modern humans left Africa, they did not sever their connection to their ancestors, just as we today have not lost our connection to those before us. It does not matter that we have become obsessed with our mid-range ancestors at the expense of thinking about our deeper ancestry; there is still an unbroken connection between the first humans and all of us. There is also an unbroken connection shared between all people alive today. It's there, no matter how intensely we may dwell on differences and divisions.

THAT DREADED "E" WORD

The last few million years of human prehistory are fascinating and important. Unfortunately, they are also controversial, due to the seemingly endless clash between science and religion. I wish I could just dodge the subject altogether, but it should be addressed, because a basic awareness of our evolutionary past can help us to understand contemporary biological diversity. Before plunging in, however, let's clear the air about evolution as far as we can. This has been a raging controversy since Charles Darwin published *On the Origin of Species* in 1859, one hundred fifty years ago. Why is evolution, the foundation of modern biology and many other modern sciences, still so contentious? Today all life sciences operate based on the conclusion that life has evolved and continues to evolve. Despite what goes on in politics and in the mass media, there is no controversy within the scientific community about whether or not life evolves.

One reason why people reject evolution is, of course, religious belief. They see it as a choice between one or the other, religion or evolution. It is not, however. I have met many people who accept evolution and still have no difficulty believing in a god or gods. Another common reason some people have difficulties with evolution is that they do not understand what it is. They may have had it explained to them by someone who was not sufficiently educated or who was intentionally misleading about it. For example, a common misconception is that the theory of evolution describes the creation of life. It does not. Darwin speculated about the origin of life, but his theory of evolution describes only how life changes. It does not explain the beginning of life. That is a separate question, one that, so far, science has not been able to answer conclusively. Therefore, if divine creation is of primary importance to you, please understand that human evolution does not necessarily contradict that belief.

Another problem may be that many people shy away from learning about evolution because they assume it is too complex. If you are one of those people, relax. Evolution is not theoretical physics. It's far easier to grasp than, say, an IRS form or the instructions that come with DVD players these days. Anyone can understand evolution.

Here are the basics: life-forms change over generations primarily due to genetic mutations and natural selection. That's it; no big deal. There are a million details, of course, but if you understand that life-forms change due to mutations and natural selection, you get the core idea of evolution.

Genetic mutations are quirks that pop up all the time in DNA. Most of them amount to nothing, while some prove harmful and others may be beneficial. Natural selection is the impact of the environment (climate, predators, parasites, prey, and so on) on life-forms. If a tiny inheritable trait provides some advantage in the environment, then those who have that trait may have more success in passing on their genes to more offspring than those who do not have it. Eventually, if that trait keeps providing an edge to enough members of the species, then it could become the norm, and that life-form will have changed, or evolved. There are many more details, but that's really all you need to understand in order to have a basic grasp of evolution.

Notice that this description of evolution did not include a declaration that all gods are nonexistent or that all religions are invalid. No matter what you may have heard, evolution does not necessarily have to conflict with, contradict, or overrule religious belief. Now if your religion demands that you believe the world is ten thousand years old and life has never changed from the instant it appeared, then, yes, you do have an irreconcilable conflict with evolution. But if that is the case, you don't just have a problem with modern biology. You also have big problems with modern botany, geology, astronomy, cosmology, microbiology, bacteriology, virology, zoology, entomology, medical science, marine biology, anthropology, ecology, and many other "–ologies." It may help some evolution deniers if they understand that there are hundreds of millions of sincere and faithful Christians, Muslims, Hindus, Buddhists, Jews, and other believers around the world who accept evolution without feeling any need to abandon their religion.

If by chance you find that you cannot or will not accept evolution, please don't think it is a deal-breaker on the issue of race. Understanding that biological races are not real is not dependent upon understanding evolution. There are plenty of problems with the concept of race that do not directly involve evolution.

THE FAMILY ALBUM

We now understand a great deal about our ancestors of the last few million years, but much remains to be discovered and reasoned out. While the overall picture remains fairly clear, discoveries continue to add new wrinkles to the story and raise even more questions. Highly respected scientists disagree on some points, and the list of named hominid species from the last few million years is long. This can all be quite confusing to the layperson, so I will spare you all but some of the most important points here.

First we must know who's who in our family album. Many fascinating species of hominids have roamed the Earth. ("Hominids" is a general term that includes modern humans and other primates who were bipedal, that is, walked on two feet.) Some of these hominid lines were evolutionary dead ends, while others were key to our existence today. This is not a complete list by any means, but at the very least, we should be aware of the following cast of characters:

Ardipithecus—older than 4 million years ago

Fossils of *Ardipithecus ramidus* were discovered in Ethiopia in 1992 but only described to the public in October of 2009. After seventeen years of analysis, the team of paleoanthropologists who found her say "Ardi" was about four feet tall (1.2 meters), spent a lot of time in trees, and was definitely bipedal on the ground. Her brain was about the size of a modern chimpanzee's or about a third the size of ours. Dated to 4.4 million years, *Ardipithecus* represents the oldest hominid fossils discovered to date. She could be one of our direct ancestors but more fossils will have to be found before we can know for sure.

Australopithecus—2 to 4 million years ago

These four-foot-tall bipedal apes lived in Africa. There were a few species of them. The famous "Lucy" was an *Australopithecus afarensis*. She was discovered by Dr. Don Johanson in Ethiopia in 1974. Australopithecines are believed to have lived in social groups but are not known to have used tools, controlled fire, or built shelters.

Homo habilis—2.5 million years ago

Homo habilis is the oldest hominid in our genus (*Homo*). The case for *Homo habilis* having human status rests on the fact that it had a larger brain size than *Australopithecus* and on the discovery of stone chopper tools that *Homo habilis* used.

Homo erectus—2 million to 300,000 years ago

What's not to love about these guys with their shallow foreheads, thick eyebrow ridges, short bodies, and sharp minds? We heap so much attention and praise on modern people who have made great discoveries and inventions. Why not *Homo erectus* too? They are thought to have been the first hominids to control fire, cook food, and build shelters. These guys probably had some form of language and they were the first hominids to walk out of Africa and reach Europe and Asia. It seems that they were not able to endure, however, as most paleoanthropologists think they became extinct a few hundred thousand years ago.

Homo floresiensis ("Hobbits")

Although *Homo floresiensis*, discovered in 2003 in Indonesia, is one of the most fascinating finds in recent years, I was reluctant to include it here because scientists are still sorting out just who these people were. Closely related to *Homo erectus* but only about three feet tall, *Homo floresiensis* appears to have been another human species that lived until as recently as eighteen thousand years ago, sharing the Earth if not their island with modern humans. Despite the fact that they had only a chimp-sized brain, some scientists believe these tiny people were capable of impressive intelligent thought. Sophisticated stone tools found alongside their fossils seem to support this.

Neanderthals—200,000 to 30,000 years ago

These stocky, muscular humans had a complex culture and lived successfully in Europe for a long time. It is important to bear in mind that modern humans did not evolve from Neanderthals. They coexisted. When biologically modern humans arrived, however, the Neanderthals faded away. There is no solid evidence to back up the idea that modern humans and Neanderthals interbred, and this suggests the Neanderthals may have been killed off by our ancestors, outdone in competition for resources, or simply unable to keep up with some environmental change. Forget the traditional image of Neanderthals as dumb and sluggish goons who couldn't keep up with modern humans. Recent research has shown that their tools were no less sophisticated than those used by *Homo sapiens sapiens* at the time.[7] As mentioned in chapter 1, Milford Wolpoff, one of the world's leading experts on Neanderthals, says the age of the Neanderthals was the last time there were biological human races. Since the Neanderthals' demise, our species has been too similar and too closely related to justify breaking it up into subspecies or races. Race believers seem to be about 30,000 years behind the times.

Modern humans—200,000 years ago to the present

We are *Homo sapiens sapiens*, the thinking species. In a relatively brief time, we successfully came to inhabit most of the world. Thanks to powerful brains, language, and a cultural emphasis on technology, we have achieved remarkable things, such as harnessing atomic power and landing on the Moon. Although the popular perception is that our differences are great, we are in reality one of the youngest and least genetically diverse species around today.

Now that we know the key players, let's take a whirlwind tour through the last few million years of the human story. All current evidence points to Africa as the continent where the earliest human ancestors arose. The case for this is overwhelming. No fossils of any hominids more than two or three million years old have ever been found outside of Africa. Every *Australopithecus* fossil discovered thus far has been found in Africa. There is no disagreement in the scientific community about Africa as the cradle of humankind. Africa is home. We are all Africans, descendants of the first African people. This is an important fact to keep in mind, particularly for those who seem to care so much about ancestry as it relates to race belief.

Try to imagine small bands of australopithecines in east Africa a few million years ago, dodging danger at every turn. Predators, disease, hunger, and perhaps even rival hominids stalked them every day. Yet they were tough enough and smart enough to run that merciless gauntlet and keep their genes moving forward one generation at a time. If those australopithecines had not succeeded, we wouldn't be here today.

It is strange that the australopithecines are unknown to most people in the world today. I thought about them during a visit to Africa. Alone and far out in the bush, I knelt at a river bank and tried to imagine these distant relatives.

I "saw" them walking along the opposite bank across the narrow river. There were about twelve of them, spread out with a few meters between each one of them. The first thing that struck me was how beautiful they were. They had a commanding presence that brought a smile to my face. Too many artists and filmmakers depict our prehistoric ancestors as filthy, tic-infested brutes. I have a different view of them.

The australopithecines I imagined that day were magnificent creatures. They appeared clean, bright, and confident. Even as they moved at a slow, casual pace, I could see obvious grace, power, and alertness. I admired them instantly and felt proud to share a connection with them.

I base this upscale view of our distant ancestors partly on my experiences of seeing animals in the wild in Africa today. During my university days, I worked as an educational aide at one of the largest zoos in the United States. I worked closely with chimpanzees, giraffes, hippos, elephants, and many other species. The animals had relatively nice habitats and were very well cared for by modern zoo standards. But they were just sad shadows of their counterparts in the wild. In Africa, I saw giraffes prancing playfully and elephants trotting with great energy. One morning I saw lions posed like magnificent marble statues in the morning sun—only these lions were fully alive, nothing like the ones I knew at the zoo. I saw troops of baboons, each animal with hair shining as if it had just been washed by professional groomers. If these animals can appear so glorious, confident, and noble in their natural environment, how might the smartest creatures in all the world three million years ago have looked and behaved?

OUT OF AFRICA—AGAIN

The "Out of Africa Model" is widely considered to be the best explanation of how prehistoric humans ended up living on six continents and many islands. This model claims that the *Homo erectus* populations that had migrated out of Africa more than a million years ago were replaced by anatomically modern humans. These modern people had evolved in Africa from *Homo erectus* or *Homo ergaster* and then made it across narrow straits in the Red Sea somewhere in the region of what is today Ethiopia and Eritrea. With little or no interbreeding, *Homo sapiens sapiens* (anatomically modern humans) replaced the *Homo erectus* or archaic *Homo sapiens* populations in Eurasia.

Amazingly, the first group of modern humans who made that monumental hike out of Africa may have numbered no more than one hundred fifty. Think about this for a moment. A small population of modern humans in Africa, probably less than thirty thousand, and an even smaller group of individuals who left Africa.[8] That's it. Together, they alone represented the sum of humanity's potential. Our entire existence rode on their shoulders.

The group that left made it to the Middle East and survived; successive generations spread into Europe, Asia, and eventually the New World (North and South America, the Caribbean). All of this happened in the last one hundred thousand years, maybe as recently as the last fifty thousand years, according to the Out of Africa Model. That's fast and very recent (remember the Cosmic Calendar). It's important to be clear that this movement of modern humans along the coasts of Eurasia was not necessarily a purposeful migration with the ambitious goal of reaching new continents. Remember, it took tens of thousands of years for these people to make it from Africa to Australia and North America. More likely it involved a series of stops and starts as people expanded further, generation after generation. Perhaps they exhausted the local resources and then moved on, maybe they were just curious, or maybe the idea of the "grass is always greener" is older than we think.

It is not known what happened when anatomically modern humans encountered the people who already had been living in Europe and Asia for thousands of years. It is possible that the *Homo erectus* populations couldn't keep up with changing environments. Perhaps the newcomers were able to outdo them in the competition for resources, thanks to better brains. Or maybe the anatomically modern humans slaughtered their cousins in some sort of prehistoric holocaust. It is possible that they interbred, but this is now thought to be unlikely, according to recent genetic research.[9]

A majority of paleoanthropologists favor the Out of Africa Model today. Science is always open to self-correction, of course, and if new evidence comes along, minds will change and textbooks will be rewritten. That's why science works so well. For now, however, the Out of Africa Model seems to be the explanation that best matches the evidence. On the issue of race, the Out of Africa Model does a good job of explaining why people today are so genetically similar. One can't overstate the significance of our genetic closeness to one another. Not only are we a very young species, but we are all very closely related. One small group of wild chimps in Africa, for example, has more genetic diversity within it than the entire human species.[10] All of the variation we see in people today—skin color, hair type, facial features, and so on—are both recent and largely superficial developments. Despite cultural differences and our tendency to focus on skin color, hair type, and facial features, we still are 99 percent alike.

The common racial identifiers that have been used to direct so much mischief and nonsense are just not that big a deal in the overall picture. Some thirty thousand years ago, some people were adapting to environments that favored darker skin and others to environments that favored lighter skin. So what? Melanin, the substance in your skin that determines

Key Points to Keep in Mind When Thinking about Race:

- Anatomically modern humans have been around for some 200,000 years;
- The first modern humans evolved in Africa. All modern humans today are descendants of these first people;
- Some modern humans migrated out of Africa between 100,000 and 50,000 years ago. Over a period of thousands of years, humans successfully spread through the Middle East, Europe, Asia, Australia, the Pacific islands, North America, South America, and the Caribbean islands;
- While living in different regions with different environmental conditions, people evolved a variety of physical traits in response to those environments. These changes include things like skin color, hair type, height, facial features, and body shape;
- Mating between people of different social groups, including "races," has been the norm throughout the human past;
- We are a relatively young and very closely related species.

color, is not tied to intelligence, morality, sexuality, athletic ability, or innate value as a human being. It determines skin color and acts as a filter for sunlight. It affects the risk of sunburn and skin cancer as well as the ability to make vitamin D. While these are important health matters, melanin just shouldn't matter as much as it still does to twenty-first-century people. We have hats, sunscreens, and sunlamps. More importantly, however, we know enough about our shared past to see that skin color, hair type, and facial features are not central to the human story. They are simply adaptations that came about along the way.

It's a shame that we don't have a few more hominid species still hanging around so that it would be easier for people to see how alike we really are. For example, if we could stand one of the Hobbits (*Homo floresiensis*) next to a few modern humans—of any color—it shouldn't be difficult to see who is more closely related to whom. Compare an *Australopithecus afarensis* and a Neanderthal and you can easily recognize two hominids who belong to different species. However, if we line up modern people—let's say a black African, a white European, and an Asian—it would be plain to see that they are of the same species, regardless of differences in color and facial features.

Biological races are just not there, at least not in any meaningful and sensible way. The spread of our biological diversity defies easy categorization. We only see races because we are taught to see them and because we make the mistake of unjustifiably magnifying trivial differences. As the human story tells us, we were squeezed through a bottleneck during prehistory and came out of it a closely related family. We have not been isolated enough to drift so far apart that we can think of ourselves as a collection of human species or distant subspecies. If anything like biological races exist at all it is only as the fading shadows of tribes and clans, small groups of people who colonized the world one footstep at a time many thousands of years ago. But those faint shadows or echoes are nowhere near evidence enough to maintain walls between millions of people. While we can be proud of carrying the genes of such capable and ingenious prehistoric people within us, we are not obligated to carry their prehistoric divisions into the future. We can—if we choose to—seek to understand our past, appreciate our diversity, and accept our closeness. Most important, we can do all this while resolving to think of ourselves as no longer belonging to tribes and clans but rather to all humankind.

Yes, race can seem very real when we watch people parade by us on a Manhattan or Paris sidewalk. It can seem obviously real when we fixate on differences and compare the physical features of people who have a greater ancestral distance between them than others. And it all seems so credible and respectable when we read or hear matter-of-fact declarations that racial groups are real. But we must look closely at every such claim, because when we ask for rock-solid evidence for the reality of biological races, invariably it is absent. The ironclad proof or the irrefutable argument never materializes. Anthropologist Marks says the idea of races became popular a few centuries ago because it was a good way to think about the world. And, he says, it still is—so long as you don't think about it too hard.[11] That is the key to race belief. It's faith without reason. It makes perfect sense—until you ask questions. Races may be convenient shorthand for describing the world. But read this shorthand closely and you will find there are far too many mistakes to give credibility to the story it tells. When challenging race belief, we must keep the human story in mind as well as every human alive today. The contrast between a dark-skinned "black" person and a light-skinned "white" person may impress our eyes but it should not impress our minds. If this contrast is exhibit A in the case for race, then the hundreds of millions of human beings of slightly different hues who stand between those two individuals are exhibit A for the case *against* race.

I have seen much of the world and many of its people. I can conceptualize no races or, if pushed, millions of races. But I cannot see how it is possible to split ourselves into a few races or several races based on any logical and consistent criteria. In the previous chapter, anthropologist Marks compared races to elusive angels. To me, biological races are more like ghosts haunting us in the age of science. Many people claim to see them but nobody ever manages to capture one for analysis and confirmation. It's your choice to believe in them or not.

Choose wisely.

NOTES

1. Ashley Montagu, *Man's Most Dangerous Myth* (Walnut Creek, CA: AltaMira, 1997), p. 126.

2. Paul Ehrlich, *Human Natures: Genes, Cultures, and the Human Prospect* (New York: Penguin Books, 2002), p. 290.

3. Steve Olson, *Mapping Human History* (Boston: Houghton Mifflin, 2002), p. 2.

4. Carl Sagan, *Billions and Billions: Thoughts on Life and Death at the Brink of the Millennium* (New York: Random House, 1997), p. 20.

5. "The Universe in One Year," Discovery Education, http://school.discoveryeducation.com/schooladventures/universe/itsawesome/cosmiccalendar/page2.html (accessed September 15, 2008).

6. Ibid.

7. University of Exeter, "New Evidence Debunks 'Stupid' Neanderthal Myth," *ScienceDaily*, August 26, 2008, http://www.sciencedaily.com/releases/2008/08/080825203924.htm (accessed September 15, 2008).

8. Nicholas Wade, *Before the Dawn: Recovering the Lost History of Our Ancestors* (New York: Penguin Books, 2006), p. 75.

9. University of Cambridge, "New Research Confirms 'Out of Africa' Theory of Human Evolution," *ScienceDaily*, May 10, 2007, http://www.sciencedaily.com/releases/2007/05/070509161829.htm (accessed September 15, 2008).

10. "Human Diversity—Go Deeper," from *Race—The Power of an Illusion*, PBS, www.pbs.org/race/000_About/002_04-background-01-11.htm (accessed September 15, 2008).

11. Jonathan Marks, "The Realities of Races," in *Is Race "Real"?* a Web forum organized by the Social Science Research Council, June 7, 2006, http://raceandgenomics.ssrc.org/Marks/ (accessed September 15, 2008).

Chapter 3

SMARTER OR JUST LUCKY?

The Negro has had just as long as the white man to develop a civilization of his own. Thousands and tens of thousands of years have passed, and the black man has not lifted his people from the darkness of Africa. If the white and black races are equal in ability, then why have they not produced equal civilizations?

—Theodore G. Bilbo,[1]
Former Mississippi senator and governor

History followed different courses for different peoples because of differences among peoples' environments, not because of biological differences among peoples themselves.

—Jared Diamond[2]

I'm slogging through a nasty blend of mud and manure. The Maasai bring their cows into the inner compound of this circular village at night to protect them from predators and thieves. While this may be wise in theory, it comes with a price. A light rain can turn a good portion of the area into something resembling a lake of diarrhea.

I plod on (nearly swimming now) toward a hut where a young boy is frantically waving at me. Once I make it to him, he points inside the hut and gestures for me to go in. I'm not sure why I should, but I enter. It's dark and smoky. A large cockroach crawls up the wall next to me. The smell of the mud/manure mixture is strong and distracting, but the thick smoke is overpowering it. I scan my surroundings but can't make out much in the shadowy haze. Finally, I see an old woman sitting against the far wall. She says something in her language and motions for me to come close. I'm not sure why I should, but I do. She looks old and unhealthy. She chatters on, but I have no idea what she is saying. She doesn't seem threatening but she doesn't seem like someone I want to spend the afternoon with, either. Just as I am about to smile and run for the exit, she grabs my shirt. Then she extends her other hand palm-up. Okay, I get it; I've seen this one many times before. It's the official international sign for "Give me money, you rich [gentleman/tourist/bastard]." By this point, I've inhaled enough smoke and I'm ready to go. So I hand her a wad of Kenyan shillings and bolt. Later, I feel creepy and uneasy about that encounter.

Many times during my travels around Africa, black people greeted me as if I were some sort of famous millionaire. Why did they automatically ask me for money? I'm not a flashy dresser and I was often alone. Obviously, my white skin marked me as a likely tourist with money to spare: nothing too sinister in that, I suppose. But I still felt uncomfortable with this weird celebrity/rich-man status. There was nothing so special about me that I should be mobbed by children, followed around by old men, and smiled at by so many young women. Why was I so fortunate as to be able to travel to Africa to enjoy the people, the scenery, and the wildlife in the first place? How come I wasn't back in Florida where I was born, begging for money from wealthy black African tourists?

Cuzco is a beautiful city that is filled with fascinating discoveries waiting for you around every corner. At an altitude of nearly eleven thousand feet, it can be tough to breathe, but it's not tough to find things of interest. Walking this South American city's back alleys and entering its old buildings inspired me to fantasize and daydream about the past more than any history book ever did. Just outside the city, I touched the stone walls of

Sacsayhuaman. I stared at the same mountains once viewed by Inca kings, and in the faces of locals I saw glimpses of the people who lived here long before. Five centuries ago did not seem so far away; the past was close here.

Unfortunately, however, that past is unsettling, as it includes terrible theft, murder, and conquest. The Spaniards Hernán Cortés and Francisco Pizarro led small forces against the vast Aztec and Inca empires and won. Mere hand-fuls of conquistadores were able to slaughter thousands of Native American warriors and seize control of two empires made up of millions of people.

A politically incorrect question comes to mind: how were a few hundred white people able to conquer millions of brown people? I know the short answer, of course. It was because the conquistadores had horses, superior weapons, and superior armor. But how did they get those advantages in the first place? It feels a bit racist to even ask the question.

While I am passing through a small village in rural Peru, a man invites me into his hut. I try my best to breathe without choking on thick smoke from an indoor cooking fire. I smile and take in the experience. Nearly half of Peru's population falls below the government's poverty line, and I'm guessing that this family is well below that mark. I notice four children sitting on the floor. Their faces are filthy and their noses crusted with old, dry mucus. The father smiles as he shows me around his tiny home. He is obvi-ously proud, but I've been around the block a few times and see that he is performing, trying his best to charm me so that I might "tip" him at the end of his little tour. I like the guy's effort, and it's not every day one gets to hang out inside an Amerindian's home high up in the Andes Mountains, so I happily give up some cash in exchange for the "cultural tour."

But, again, why was I the one giving money to him? Why wasn't he visiting my little smoke-choked hut back in the United States or Europe, feeling sorry for my kids, taking photos of my impoverished neighborhood with an expensive camera, and handing me some of his pocket change?

Australian Aborigines probably seem like a happy bunch to people who have never visited Australia or read about that nation's history. How could they not be, blowing on their didgeridoos and posing for photos, sharing ad space with koalas and kangaroos in tourism marketing campaigns? Sadly, the reality is less charming. Australia's indigenous people suffered horrendous treatment from the white people who took control of the land they had lived on for tens of thousands of years. They were murdered and enslaved. Many "mixed" Aboriginal children were taken from their mothers. Worst of all, when the first waves of white people landed in Australia, they brought with them European diseases such as smallpox, which the indigenous people had never been exposed to. Perhaps half or more of the indigenous population perished in the eighteenth century as a result of the arrival of white people. Today things are better, of course, but still well short of ideal. Compared with white Australians, Aborigines fall short in virtually every measure of social well-being, such as income, employment, education, health, and life expectancy.

During my time in Australia, I was fortunate enough to catch a show by a brilliant Aborigine didgeridoo player in Queensland. I was swept away by the performance. Watching him puff and blow on the wooden instrument to create unique and chilling sounds allowed me momentarily to forget the past and present pain suffered by his people. That man and his music provided a pleasant break from thoughts about the madness of prejudice and conquest. On the way back to my hotel, however, my white bus driver's sense of humor ended my dreamtime. We drove by a run-down neighborhood where some local Aborigines lived, maybe even the guy who had played for me earlier. The small homes were in terrible shape. Junk was everywhere in the front yards. A rusty, broken-down car was abandoned on the roadside. I asked the driver for his take on the state of Aborigines today. Call me naïve, but I expected a compassionate response flavored with wise insight from a local. What I got was a juvenile racist joke.

"Have you heard about the special bungee jumping offer we have for Aborigines?" he asked. "It's a really great deal—no strings attached." Laughing, he added, "Yeah, that'd do the trick!"

Given what I knew about the history of white people lynching black people in America, I didn't find anything funny about the joke. Question: How did that well-fed white bus driver get to be the one driving in an air-conditioned vehicle while telling racist jokes about Aboriginals to other white people? Why didn't it end up the other way around?

I visited my hotel's gift shop during a stay at the Grand Canyon in Arizona and found that most of the merchandise had a Native American theme.

Much of it was standard tourism fodder, to feed the herd: key chains, mugs, T-shirts, and so on. Some of it, however, was very impressive, high-quality art produced by Native Americans in the area. The kachina dolls, for example, were outstanding—pricey, but outstanding. These traditional wooden figures are still carved and painted by the Hopi people, many of whom now live on reservation lands in Arizona. I watched white tourist after white tourist pause to admire the beautiful figures. I hope the Hopi artists receive a fair percentage from each sale because, based on my observations, the dolls seem to sell well.

Why were white people able to conquer Native Americans anyway? Why didn't the Native Americans crush the outnumbered invaders and push them back into the sea every time their ships attempted to land in the New World? Why didn't Native Americans beat Columbus and the Europeans to the punch and launch an invasion of Europe? These are not trivial questions, of interest only to alternative history fans. These are important questions that have profound implications for our view of race today. The traditional racist view has been that the scorecard of history is precisely what we should expect when races come into contact, because white people are superior to black and brown people, for example. But what about people who do not think in racist ways and do not hate or rank other people according to racial affiliation? How do they explain the obvious fact that Europeans, white people, have been so successful at colonizing, exploiting, robbing, enslaving, and murdering nonwhite people for the last one thousand years? The quick answer is that usually they just don't explain it. They duck and wait for the next question. White domination in most cases of historical contact between "races" is an awkward question best left unasked and unanswered in the minds of most polite people today.

I recall privately struggling with this simple question as far back as my childhood. If everyone is equal and no one race is better than another, I would wonder, then why did white people invent so many things, own millions of slaves, and conquer so much land around the world? Racism might explain why white people get good jobs or have an easier time renting apartments and catching cabs, but racism can't explain how white people were able to steal three continents out from under indigenous populations. It frustrated me, because the cold, harsh reality of history didn't seem to match up very well with what I was hearing from the adults who were preaching their sermon of racial equality. Even more confusing, white European success seemed a perfect match for the worst racist statements of all: "Of course white people are smarter than other races of people, and the proof is our wealth, military power, scientific achievements, and domination of so many nonwhite societies over the last several centuries." I have heard one version or another of that statement many times in my life.

A QUICK REALITY CHECK

Before we explore this topic further, a brief reality check is necessary. Let's acknowledge that there is a time bias at work here. We are alive now, so we tend to think that this is the most important time ever and that the last thousand years or so are critically important because they set the stage for "now." This is subjective thinking, understandable perhaps, but subjective nonetheless. When we recognize this time bias, it becomes a little less easy to believe in something like "historical domination by the white race." What justifies taking the position that only one particular portion of the past has more value than other, much larger, portions of the past? If we try to look at the past objectively, then we have to admit that white people have not dominated. In fact, white people have dominated only a very tiny fraction of human existence. For example, Africans, probably dark-skinned, have been the rulers of our species far longer than Europeans or anyone else can claim to be. Africans did not just dominate humankind; they *were* humankind.

Homo habilis, the first recognized member of our genus (*Homo*), arose in Africa more than two million years ago. According to current evidence, for about a million years the only people on Earth were the African *Homo habilis* and then the African *Homo erectus*. Then, anatomically modern humans (our big-brained direct ancestors) evolved in Africa some two hundred thousand years ago but didn't spread out of Africa until perhaps one hundred thousand to fifty thousand years ago. This means that that for at least one hundred thousand years, the mentally and culturally superior *Homo sapiens sapiens* was found only in Africa. Therefore, African people were the elite, the best, the intellectual giants of the Earth, and by far the most advanced "race," if you want to use that term. When these superior people left Africa and encountered other humans (*Homo erectus* populations that had migrated from Africa earlier), they apparently defeated them thanks to superior intelligence. Therefore, the story of modern humanity, in the context of our full two hundred thousand years, is primarily a story of African success and domination. Things have gone exceedingly well for white European people only in the last thousand years or so. That's about *half of 1 percent* of the human span of existence.

It also bears mentioning that the idea of white people's success is not as clear-cut as it may seem, not when so many nonwhite people contributed and contribute so much to the success of "white societies." Is it really accurate, for example, to think of the United States or the European nations as white success stories, when nonwhite people have worked and still work in their fields and factories, fight in their armies, run businesses, and serve in high government office? It's like saying the Boston Celtics basketball team of the 1980s was a great white team (as some fans did). But key contributions from Dennis Johnson and Robert Parish (black players) make such a claim laughable.

There are many places around the world where the underlying force that shaped the social landscape was the historical domination and/or extermination of nonwhite people by white people. But, of course, it is neither honest nor accurate to view the crimes of these white people in a racial context only. Ultimately, slavery, genocide, theft, and exploitation are about humans abusing humans. The race concept obviously plays a prominent role in these events, but it is not the only or even the most important factor. There is no reason to assume, for example, that black Africans or brown South Americans or anyone else would have acted differently than white Europeans did if these black or brown people had enjoyed a technological edge over white people. White people did not introduce the concept of slavery to black Africans; it was already there. Pizarro and Cortés did not introduce social inequality and violence to the Inca and Aztec empires.

It is so easy, too easy, to view our history as a series of collisions between bad people and good people. Unfortunately, the past does not sort out so conveniently. What is clear is that when one group has had superior social sophistication and technology, it has usually taken advantage of weaker groups. And so again we return to that nagging, important question: why didn't black and brown people have the superior social sophistication and technological advantage when the Old and New Worlds came into contact? It's not just an obvious question; it's a fair question and one that should be answered or at least discussed. Not talking about it only allows committed racists to continue unopposed in telling everyone that history is not only proof of the existence of racial groups but also proof that the white race is supposed to be on top of the pile. Not confronting this question also condemns many thoughtful antiracists to endure nagging suspicions that they just might be wrong.

The lopsided racial contest that seems to be our recent history can prompt those who do not hold a racist worldview to doubt their position, based solely on the prominence of white versus black/brown clashes of the past. If races aren't real, if racists' philosophies are wrong, then why does much of the last thousand years seem to be about various races competing and fighting, with white people faring so well most of the time?

Fortunately, a brilliant scientist named Jared Diamond was intrigued by this question and decided to make an attempt at answering it. The result of his efforts was a Pulitzer Prize–winning book, *Guns, Germs, and Steel*. Diamond, a UCLA professor of physiology and evolutionary biology and a winner of the MacArthur Foundation genius award, has conducted extensive field research in Papua New Guinea where he lived with local people. It was an encounter back in 1972 with one New Guinean, a man named Yali, that sparked Diamond's quest to find out why white people triumphed over black and brown people with such apparent ease.

"Why is it that you white people developed so much cargo and brought it to New Guinea," Yali asked. "but we black people had little cargo of our own?"

By "cargo," Yali meant all the things white people had brought to his land that New Guineans had never had before, items such as steel axes, medicine, radios, clothes, and guns. Diamond was stumped and couldn't come up with a good answer for Yali then. Some twenty-five years later, however, he had a response. It came in the form of a 480-page book. And what an answer it is.

Long ago, Diamond had looked into the race concept and had come away convinced that there is little or nothing to it. He could find no convincing case to be made for the idea that biological races are sensible categories or that races can be ranked by average intelligence. For his conclusions he relied not only upon the accumulated scientific research of the last couple of centuries but upon personal experience as well. Diamond is convinced that people living a "primitive" or "Stone Age" lifestyle today are on average more intelligent than people living in North American and European cities. He writes in *Guns, Germs, and Steel*:

> My perspective on this controversy comes from 33 years of working with New Guineans in their own intact societies. From the very beginning of my work with New Guineans, they impressed me as being on the average more intelligent, more alert, more expressive, and more interested in things and people around them than the average European or American is. At some tasks that one might reasonably suppose to reflect aspects of brain function, such as the ability to form a mental map of unfamiliar surroundings, they appear considerably more adept than Westerners. Of course, New Guineans tend to perform poorly at tasks that Westerners have been trained to perform since childhood and that New Guineans have not. Hence when unschooled New Guineans from remote villages visit towns, they look stupid to Westerners. Conversely, I am constantly aware of how stupid I look to New Guineans when I'm with them in the jungle, displaying my incompetence at simple tasks (such as following a jungle trail or erecting a shelter) at which New Guineans have been trained since childhood and I have not.[3]

Diamond offers a personal guess as to why so many New Guineans he encounters in the jungle seem more alert and engaging than many of the urban people he interacts with in the United States and Europe. He cites the fact that Europeans have been living for thousands of years in densely populated societies, where the major cause of death has been infectious epidemic diseases. Despite common perceptions, war and murder have not claimed, relatively speaking, a large number of lives. "In contrast, New Guineans have been living in societies where human numbers were too low for epidemic diseases of high populations to evolve. Instead, traditional New Guineans suffered high mortality from murder, chronic tribal warfare, accidents, and problems procuring food. Intelligent people are likelier than less intelligent ones to escape those causes of high mortality in traditional New Guinea societies. However, the differential mortality from epidemic diseases in traditional European societies had little to do with intelligence, and instead involved genetic resistance dependent on details of body chemistry."[4]

Diamond also points out that most modern American and European children spend a large portion of their time sitting in front of televisions, while traditional New Guinea children play and interact with adults and other children, engaging in activities more likely to stimulate intellectual development. In the end, Diamond concludes that there is nothing readily apparent in the condition of people today that provides an answer to Yali's question. So, he has decided, the answer must lie somewhere in the past.

Before plunging into a review of his answer to the question of why white societies have fared so well over the last several centuries, however, I want to mention that I agree at least to a degree with Diamond's general view of people in industrialized cultures versus people in preindustrial cultures. After traveling extensively on six continents, meeting and speaking with many people in a variety of societies, I also believe that there is no readily detectable generalized intellectual inferiority among the brown-skinned and black-skinned people of South America, the Caribbean, Africa, Australia, Fiji, or Papua New Guinea. Everywhere I have been, I have encountered what seemed to me to be very bright people, average people, and people who shouldn't be allowed to operate heavy machinery under any circumstances—just about the same sort of mix of people one encounters in any North American or European city. As for Diamond's view that modern "Stone Age" people are on average *smarter* than industrialized people, I can't say. I haven't spent time with people who are living a virtually Stone Age existence, so I don't know. But I have encountered rural people in Africa, Nepal, Papua New Guinea, Ecuador, and Peru who are closer to the Stone Age than the Space Age. I have no reason to think that they are smarter than people I encounter in New York City, London, and Paris. It's possible, but I can't be sure. What I can say is that I have never noticed a recognizable pattern of dim dark-skinned people and bright light-skinned people. It has been my experience that both bright people and those who are noticeably less so come in all colors.

One of the standard assumptions of the racist view of human intelligence is that different races evolved at different rates on different continents because of different environments. The late anthropologist Ashley Montagu rejected this idea, however, because it omits the impact of culture:

> While physical environments have varied considerably, the cultural environments of man during the whole of his evolutionary history, right up to the recent period, have been fundamentally alike, namely, that of food-gathering hunters.... The cultural differences between a Congo

Pygmy and an Eskimo of the Far North are only superficially different. … There are great differences in detail between the food-gathering cultures, but in the responses they have made to their particular environments they are strikingly and fundamentally alike. They are alike in their material culture, in their social organization, in the absence of chieftains, political or state organization, permanent councils, belief in gods, warfare, and the like."[5]

Diamond knows, like historians and just about everyone else, that Europeans killed and conquered nonwhite people because they had key advantages. The primary advantages, according to Diamond, were these: European guns, infectious diseases, steel tools, and manufactured products. But while this explains how Europeans were able to conquer the New World and Australia as well as enslave millions of Africans, it does not provide an explanation of *why* they had attained such a superior position well before they launched the first slave ship or planted the first flag on some distant shore. Europeans' conquests were so one-sided that they cannot be explained away as the good fortune of winning a key battle or two or the fruits of a few inventions. There must have been a big reason underlying it all. It is vital that we understand this reason, Diamond believes, because without a satisfactory explanation for white domination, most people will settle for the racist conclusion and believe that white people must be inherently smarter, comprising the superior race.

We also have to wonder about Asia. Why did many Asian societies develop highly complex cities and have impressive technology? Their success over the last several thousand years has a lot to do with the current racist belief held by many people that Asians rank near or even above the white race in natural intelligence and well above the world's black and brown people. So we must seek an answer not just for the white Europeans but for the Asians as well.

"It seems logical to suppose that history's pattern reflects innate differences among people themselves," Diamond writes:

Of course, we're taught that it's not polite to say so in public. We read of technical studies claiming to demonstrate inborn differences, and we also read rebuttals claiming that those studies suffer from technical flaws. We see in our daily lives that some of the conquered peoples continue to form an underclass, centuries after the conquests [and] slave imports took place. We're told that this too is to be attributed not to any biolog-ical shortcomings but to social disadvantages and limited opportunities.

Nevertheless, we have to wonder. We keep seeing all those glaring, persistent differences in people's status. We're assured that the seemingly transparent biological explanation for the world's inequalities as of A.D. 1500 is wrong, but we're not told what the correct explanation is. Until we have some convincing, detailed agreed-upon explanation for the broad pattern of history, most people will continue to suspect that the racist biological explanation is correct after all.[6]

What, then, is Diamond's explanation for the white man's ability to bully so many people around for the last five centuries or so? Surprisingly, Diamond found his answer in the plants and animals available for domes-tication by the first Eurasians. The reason why Europeans were able to become such successful conquerors, enslavers, and colonizers, Diamond concludes, was that they had natural resources that people in Australia and the New World did not. Diamond's case hinges on the natural presence or absence of plants and animals that were relatively easy to domesticate. As it turns out, people in the Middle East and Eurasia twenty thousand years ago just happened to have the good fortune of living in environments that were ideal for building complex civilizations. Meanwhile, people in Africa, Australia, and the New World had very little to work with, by comparison. This could be nothing more than a big coincidence, of course, but it does seem to explain a lot.

THE POWER OF PLANTS

The achievement of consistent, large-scale food production is a challenge that must be met before a society can become a complex urban culture. Hunting and gathering simply won't do it. A small band of humans who live off what the land offers naturally can never settle down and establish a booming metropolis with a population of many thousands or millions of inhabitants. It just can't happen.

The necessary first step is to domesticate plants and animals, to learn how to control them, change them, and exploit them. Anthropologists tell me that this process probably was not due to a series of "Eureka!" moments experienced by individual prehistoric geniuses, but more likely it was a long process in which succeeding generations observed, experimented, and, through trial and error, figured out which plants and animals could permanently be put into service for people. Once that process is underway and delivering sufficient dividends, people can stop moving. They can settle down in one place to plant and harvest crops and to raise livestock. This, of course, sets the stage for a full-blown civilization to grow. Agriculture provides enough food for larger and larger stationary populations, which in turn allows for complex societies in which not all the people are needed to procure food. Some people can now specialize in and devote their time to construction, engineering, crafts, arts, religion, warfare, and government, for example. The domestication of plants and animals is a crucial launching point for civilization and all the things, good and bad, that come with it. So why did some people make that leap sooner and more successfully than other people? Were they smarter? Or could they simply have been lucky enough to be surrounded by the best resources, the ideal wild grasses and animals? Diamond thinks it was the latter.

Consider the distribution of plants. With more than two hundred thousand wild flowering plants around the world, one might think that any prehistoric people living in any moderate climate zone could easily have selected a few species to domesticate and been off to the races. It wasn't anywhere near that simple, however. Diamond explains that the vast majority of wild plants are inedible, only a few thousand are eaten by humans, and, of those, only a few hundred have been domesticated. But the numbers go way down when you toss out the minor contributors. It turns out that that just a dozen species contribute more than 80 percent of the world's annual crops. They are the cereals (such as wheat, corn, rice), the soybean (a pulse), the roots or tubers (such as the potato), the sugar sources (sugar cane, sugar beet), and the banana (a fruit). "With so few major crops in the world, all of them domesticated thousands of years ago," Diamond explains, "it's less surprising that many areas of the world had no wild native plants at all of outstanding potential. Our failure to domesticate even a single major new food plant in modern times suggests that ancient people really may have explored virtually all useful wild plants and domesticated all the ones worth domesticating."[7]

Since the people of Europe and Asia were the first and the most successful at domesticating plants and animals—giving them a crucial head start toward complex and technologically powerful societies—can we conclude that this means Eurasians were smarter than, say, Australian Aborigines, who never made that leap? No, argues Diamond, because the deck was stacked heavily in favor of the Eurasians. The wild plants and animals available to Eurasians comprised a far superior pool from which to domesticate useful species. Diamond cites, for example, the research of Mark Blumler, who surveyed the global distribution of large-seeded cereals or grass species, described by Diamond as "the cream of nature's crop." Blumler found that of the world's fifty-six large-seeded species, thirty-three were in the hot spots of early agriculture and civilization: West Asia, Europe, and North Africa. How many did sub-Saharan Africa potentially have to work with? Only four. All of North America had just four as well. Mesoamerica had five and South America a mere two. Northern Australia had two, while southwestern Australia had none.[8] "That fact alone goes a long way toward explaining the course of human history," concludes Diamond.[9] It is not unreasonable to believe that people who lived among thirty-three species of wild grasses with a high potential for human use enjoyed a decisive advantage in developing agriculture and thus civilization. It also seems reasonable to believe that people who had few or none of these large-seed species faced a significant disadvantage, no matter how smart they may have been.

Natural Distribution of the World's 56 Large-Seeded Species That Were the Best Candidates for Domestication[10]

Sub-Saharan Africa	4
North America	4
Mesoamerica	5
South America	2
Northern Australia	2
Southwestern Australia	0
Eurasia, North Africa	33

THE ANIMAL ADVANTAGE

What about domesticated animals? Why were Eurasians so far ahead of everyone else in taming and exploiting animals? Remember, this too was a crucial step toward civilization, the division of labor, and the development of government bureaucracies and military power. Why were European conquistadores on horseback rather than Native Americans? Was it the product of a racial mental advantage or something else? Once again, we find that there were significant differences in the natural resources available to people living on various continents prior to the rise of civilizations thirteen thousand years ago. The reason Eurasians had so much success at domesticating animals such as pigs, horses, cows, sheep, and goats doesn't seem to have been dependent upon Eurasians being more intelligent than sub–Saharan Africans, Australians, and Native Americans. More likely, the reason was simply that they had many more opportunities. Diamond shows that the distribution of large mammals that were good candidates for domestication heavily favored Eurasians. This should not be surprising, because Eurasia is bigger and more ecologically diverse. By "good candidates," Diamond means any species of land-dwelling herbivore or omnivore that weighs on average more than one hundred pounds. Nature's global distribution was anything but fair. Eurasia had seventy-two animal candidates for domestication, and Eurasians successfully domesticated thirteen of them (that's a rate of 18 percent). Sub–Saharan Africa had fifty-one candidates. None were domesticated. The Americas had twenty-four candidates for domestication but only one, the llama/alpaca, was domesticated (a rate of 4 percent). Don't be misled by mental images of Native Americans riding on horseback in the Wild West many years ago. Horses were imported into the Americas by early European explorers and settlers. Australia's lone candidate was not domesticated.[11]

Animal Candidates for Domestication[12]

	Eurasia	Sub-Saharan Africa	The Americas	Australia
Candidates	72	51	24	1
Domesticated species	13	0	1	0
Percentage of candidates domesticated	18%	0%	4%	0%

So, out of all the world's reasonable candidates for domestication thirteen thousand years ago, Eurasians domesticated thirteen of them and Native Americans domesticated one. Anyone who does not know about this original distribution of wild animals might naturally drift toward some notion of "racial" intelligence as the best explanation for such drastically different rates of domestication. Awareness of the facts, however, changes everything. Obviously Europeans could not have matched their success in domesticating animals if they had found themselves not in Europe thirteen thousand years ago but in Australia, for example. They would have had so little to work with that it would have been impossible. On the other hand, it doesn't take too large a leap of imagination to see Australian Aborigines doing quite well at making use of sheep, cows, pigs, and horses if they had lived in Eurasia thirteen thousand years ago.

You may be wondering why sub-Saharan Africa had a complete shutout. The region had fifty-one candidates. That's a lot. Why weren't any of them domesticated? Why didn't black Africans ride zebras and make rhinoceroses pull their plows? Diamond argues that it has nothing to do with human intelligence. Sub-Saharan Africa had the wrong animals for domestication. If the people there had had access to more agreeable animals, they almost surely would have exploited them. A relevant fact is that some of the animals that were domesticated by Eurasians were domesticated independently by different peoples in different regions of the Eurasian landmass. This suggests that domestication was not dependent upon some extraordinary level of intelligence unique to a particular people or community. It most likely had everything to do with the right species simply living near human communities. Another important point is that people everywhere, including those in sub-Saharan Africa, readily accepted and utilized Eurasian domesticated animals when they became available. There were no mental obstacles to this, as Diamond makes clear:

> Surely, if some local wild mammal species of those continents had been domesticable, some Australian, American, and African peoples would have domesticated them and gained great advantage from them, just as they benefited from the Eurasian domestic animals that they immediately adopted when those became available. For instance, consider all the peoples of sub-Saharan Africa living within range of wild zebras and buffalo. Why wasn't there at least one African hunter-gatherer tribe that domesti-

cated those zebras and buffalo and that thereby gained sway over other Africans, without having to await the arrival of Eurasian horses and cattle? All these facts indicate that the explanation for the lack of native mammal domestication outside Eurasia lay with the locally available wild mammals themselves, not with the local peoples.[13]

It is also important to be aware that simply taming an individual animal is very different from domesticating a species. When a species becomes thoroughly domesticated, that usually means it has developed a natural predisposition to human captivity. The whole species or subspecies itself has been changed, not just an individual animal. Diamond explains that many wild animals may be tamed (giraffes, elephants, even hyenas) but only a few species have ended up serving humans as fully domesticated species. Even in modern times, it has proven difficult if not impossible to domesticate many large mammalian species.[14] It appears that some species of wild animals were highly conducive to domestication and some, it seems, may never be.

THE SHAPE OF THINGS TO COME

Another crucial advantage Eurasia had over sub-Saharan Africa and the Americas was shape and orientation. It is important to understand that not all societies in Europe and Asia simultaneously became complex, agriculture-based city-states. Only a few people in a few places made that transition in-dependently. However, other Eurasian people followed them in short order. This was not the case in Africa or the Americas. Why?

The most likely reason is the fact that the Eurasian supercontinent was shaped and oriented in a way that was highly conducive to the rapid spread of new ideas about agriculture and animal domestication. Communication and trade were easy in much of Europe and Asia and, no less important, many people were living at roughly the same latitudes, which means they shared similar environmental conditions such as temperature, length of day, and seasons. This is crucial for people who might want to share information on farming and animal care or actually trade seeds and animals. According to Diamond's research, a shared plant species is more likely to do well even four thousand miles away—if it remains at roughly the same latitude. However, if the plant species is even as little as one thousand miles south or north of its natural environment, the odds of success drop significantly. "Woe betide the plant whose genetic program is mismatched to the latitude of the field in which it is planted!" warns Diamond.[15]

Furthermore, it's not just seeds that are picky about latitude. Domesticated animals that do well in one latitudinal range have a much better chance of faring well at a similar latitudinal range, even if it is many thousands of miles to the east or west.

"That's part of the reason why Fertile Crescent [Southwest Asian] domesticates spread west and east so rapidly," explains Diamond. "They were already well adapted to the climates of the regions to which they were spreading. For instance, once farming crossed from the plains of Hungary into central Europe around 5,400 BC, it spread so quickly that the sites of the first farmers in the vast area from Poland west to Holland were nearly contemporaneous. By the time of Christ, cereals of fertile crescent origin were growing over the ten-thousand-mile expanse from the Atlantic coast of Ireland to the Pacific coast of Japan."[16] The side-to-side orientation of Eurasia not only allowed agriculture and animal domestication to spread, but it also facilitated the relatively easy and rapid spread of new ideas and technologies. This was yet another huge advantage for Europeans and Asians. Consider the Americas, by contrast. Some nine thousand miles from north to south, the two continents are just three thousand miles across at their widest and only forty miles wide at their narrowest point. It was simply never in the cards for people who might have domesticated a wild plant in what is now Maryland, for example, to easily share that success, to transfer it to people living in what is now Brazil or Argentina. And even if they did, the odds are that the particular crop would have failed in its new environment. Because the shape and orientation of the Americas was not helpful to the exchange of the ideas and technology of food production between prehistoric societies, the people living there were at a clear disadvantage. Compared to Eurasians, they had a much more difficult path toward developing large, complex societies based on plant and animal domestication.[17]

Given the different circumstances faced by people on different continents, the idea of inherited intelligence seems less likely to be a valid explanation for the different rates at which complex societies arose. Geography and environment seem to be the basic, sensible explanations. Inheriting a favorable continent is not the same as inheriting superior intelligence.

SUB-SAHARAN AFRICA

Like the Americas, sub-Saharan Africa faced geographical challenges that Europeans were spared. Consider Egypt, one of history's earliest and most successful civilizations. This nonwhite African society produced one of the world's earliest and greatest civilizations. But two facts cannot be overlooked: Egypt is north of the Sahara Desert and was, therefore, connected to the Mediterranean's regional information and trade network. It was also situated in a climate zone similar to that of Southwest Asia (the Fertile Crescent).[18] So, it's likely that the glory of Ancient Egypt was made possible by the spread of ideas, technology, and seeds across the accommodating geography of Eurasia and northern Africa. But while North African societies benefited from that Eurasian highway of trade and talk, societies in sub-Saharan Africa did not. The spread of both domesticated plants and animals from Eurasia was impeded, if not stopped, by that great roadblock named the Sahara Desert as well as by middle Africa's environmental conditions, according to Diamond:

> South Africa's Mediterranean climate would have been ideal for [domesticated plants and animals], but the 2,000 miles of tropical conditions between Ethiopia and South Africa posed an insurmountable barrier. Instead, African agriculture south of the Sahara would have to be launched by the domestication of wild plants (such as sorghum and African yams) indigenous to the Sahel zone and to tropical West Africa, and adapted to the warm temperatures, summer rains, and relatively constant day lengths of those low latitudes.... Similarly, the spread southward of Fertile Crescent domestic animals through Africa was stopped or slowed by climate and disease.... The horse never became established farther south than West Africa's kingdoms north of the equator. The advance of cattle, sheep, and goats halted for 2,000 years at the northern edge of the Serengeti Plains, while new types of human economies and livestock breeds were being developed. Not until the period A.D. 1–200, some 8,000 years after livestock were domesticated in the Fertile Crescent, did cattle, sheep, and goats finally reach South Africa.[19]

An interesting detail mentioned in *Guns, Germs, and Steel* is that most Fertile Crescent crops can be traced genetically to a single event or process of domestication. Once a new crop worked well, the news, and the seeds, spread so fast throughout Eurasia that there was no need for other people to repeat the domestication process. They just copied their neighbors. Contrast this with the Americas, where many similar wild plant species were domesticated independently by different people in different places.[20] An alternative racist view of history might point to this as evidence that Native Americans were more inventive and had greater natural intelligence than Europeans and Asians!

BIOLOGICAL WARFARE

There is another big reason why Europeans were able to conquer so many people in so many faraway lands with relative ease. That reason is germs. Deadly diseases have always been with us, helping to shape human prehistory and history every step of the way. This was the case when the Old World and the New World collided five centuries ago. In that instance, germs helped the white Europeans. The Europeans did not understand germs and disease at that time so they cannot be blamed for their role in this horrendous biological war; nor can anyone sensibly cite this in support of the claim that white Europeans were inherently smarter than the people they conquered.

Few people are aware of the power and influence of germs. These little creatures all around us, on us, and inside us—were the first forms of life on this planet and will almost certainly be the last. The germs and microbes on your skin and in your body right now outnumber your own cells. You are mostly an ecosystem made up of "other" life-forms. One can make a very strong case that they, not us, are the most successful and dominant life-forms on Earth today. Microbes and germs inhabit virtually every environment. They live in ice, rock, and super-heated water.

They also make history.

Many times, the course of human events has been influenced more by germs than by kings, generals, and inventors. Political, religious, social, and economic institutions were shaken up, for example, when the Black Death ravaged Europe, killing some 50 percent of the population. Nobody gave them any medals, but germs killed more soldiers in World War I (and most other wars) than bullets and bombs. And European germs killed far more Australian Aborigines and Native Americans than European guns and steel swords did.

From the moment Europeans landed in the New World and in Australia, an unintentional though ferocious biological war was waged. While most of us have little difficulty imagining Europeans stabbing, shooting, and enslaving the people they encountered, it is not so easy to wrap one's mind around tens of millions of indigenous people being killed by invisible germs delivered via European ships. But it really did happen, and it may have been the most important factor of all in the Old World's conquest of the New World. Again, Europeans cannot be blamed for this aspect of their brutal assaults on indigenous peoples. There were no microbiologists and epidemiologists around in the fifteenth century, so it is unlikely that early explorers understood beforehand that their mere breath and touch could cause such a massive death toll among native people. So why did Europeans end up being the ones who could carry but survive infectious germs, while so many native people in the Americas and Australia could not?

One thing is certain: Europeans' development of a general resistance to smallpox, influenza, measles, and other potentially deadly diseases had nothing to do with anything that can be called racial intelligence. It came about because they lived in close proximity to those diseases for thousands of years. The germs that killed so many people in the New World and in Australia originally made the jump to Europeans from the animals that Europeans had domesticated. When people began working and living with goats, pigs, and cows on a large scale, an endless siege began. Germs crossed over and repeatedly ravaged densely populated European societies. This allowed (forced) human populations in Europe to evolve with genetic immunities. The result was that European people evolved to have immunities and most people were able to survive, whereas populations in the Americas and Australia did not have that experience of thousands of years with close exposure to the animal sources of these germs and therefore never evolved the capacity to defend themselves.

Do not overlook the impact of germs. Some estimates of the lives lost to European diseases in the New World are as high as 95 percent of the population. If this is anywhere near the real death rate, then this terrible event was not so far from the total annihilation of all native peoples in North America, South America, Central America, and the Caribbean islands. There is no way ever to know, but obviously this could have been the most important factor of all in the success of the Europeans in the

New World. Maybe superior weapons, military tactics, and organization would not have been enough if inadvertent germ warfare had not conveniently decimated most of the indigenous population. Again, it seems white people were not smarter—just luckier.

LOCATION, LOCATION, LOCATION

The key point Diamond makes in *Guns, Germs, and Steel* is that the critical reasons for European success seem to have had everything to do with real estate and nothing to do with inherited intelligence based on racial groups. Europeans weren't smarter than Africans, Australians, and Native Americans. It seems they had the good fortune to be born in lands with superior advantages and opportunities. Eurasians were first and best at domesticating plants and animals, not because they had better minds but because they had better plants and animals to exploit. Good luck, not a racial intelligence advantage, put Eurasia on the fast track to civilization and military superiority. Bad luck, not a racial intelligence disadvantage, put the people of Australia, Africa, and the Americas behind and left them vulnerable to European weapons and germs. Ultimately, Diamond is able to answer Yali's question in a mere twenty words: "History followed different courses for different peoples because of differences among peoples' environments, not because of differences in peoples themselves."[21]

I have a university degree in history and I am well aware that professional historians cringe at simple answers to complex questions. Because of this, I read *Guns, Germs, and Steel* with a skeptical eye. By the time I closed the book, however, I was convinced that Diamond had made a compelling case. Sure, there is more to the story—there always is—but it is clear that environmental conditions several thousand years ago offer a much better explanation for history's scorecard than the idea that some continental populations or "races" are inherently more intelligent than others.

I believe it is important for as many people as possible to be made aware of Diamond's ideas in *Guns, Germs, and Steel*. Diamond's work does not prove that biological races do not exist; nor does it prove that there is no link between intelligence and race. It does, however, provide a credible alternative explanation for the lopsided results of history. No longer does anyone have to side with the traditional racist interpretation of history by default, because there is nothing else on the table. Nonracist people do not have to remain silent or change the subject when someone asks why Europeans and Asians did so well in comparison with other peoples over the last thousand years or so. There is an answer. Diamond's work may not be the complete answer, and it may one day be dismantled and defeated by better ideas. For now, however, it makes a lot more sense than the oft-heard assertion that the white race is inherently more intelligent than other races.

NOTES

1. Theodore G. Bilbo, Take Your Choice: Separation or Mongrelization, www.churchoftrueisrael.com/tyc/tyc-06.html (accessed July 19, 2009).

2. Jared M. Diamond, *Guns, Germs, and Steel: The Fates of Human Society* (New York: Norton, 2005), p. 25.

3. Ibid., p. 20.

4. Ibid., p. 21.

5. Ashley Montagu, *Race and IQ* (New York: Oxford University Press, 1999), pp. 11–12.

6. Diamond, *Guns, Germs, and Steel*, p. 25.

7. Ibid., pp. 132–33.

8. Ibid., p. 140.

9. Ibid., p. 139.

10. Ibid., p. 140.

11. Ibid., p. 162.

12. Ibid.

13. Ibid., p. 164.

14. Ibid., pp. 163–74.

15. Ibid., p. 184.

16. Ibid., p. 185.

17. Ibid., pp. 176–77.

18. Ibid., p. 182.

19. Ibid., p. 186.

20. Ibid., p. 188.

21. Ibid., p. 25.

Chapter 4

BLACK MEN CAN'T JUMP

Men occasionally stumble over the truth, but most of them pick themselves up and hurry off as if nothing has happened.

—Winston Churchill[1]

Race is the witchcraft, the demonology of our time, the means by which we exorcise imagined demoniacal powers among us.

—Ashley Montagu[2]

Here's something interesting about race and sports that you may not be aware of: black men can't jump. That's right: when it comes to leaping ability, black people are positively lead footed compared to white people. That 1992 film, *White Men Can't Jump*, starring Wesley Snipes and Woody Harrelson, lied to you. Proof of this can be found in the objective results of international sports competitions. If one wants to determine who the best jumpers are, where better to look than high jump competitions, events participated in by virtually all races and countries? The results speak for themselves. Seven of the ten best high jumpers of all time (based on height cleared) are white. Only three are black.[3] A review of the Track and Field News World Rankings for this event from 1947 through 2002 also shows clear white domination. These rankings are widely respected and cited by knowledgeable people within the track and field community. They are created by votes from experts who base their selections on marks, honors won, and win-loss record for the season. Although black athletes have topped the rankings in some years, the overall pattern is clear. White men are the world's best high jumpers. A review of the world rankings for women (1956–2002) turns up the same pattern. White women have been the best female high jumpers by far.[4]

Nothing appears to be likely to change any time soon. Only four black athletes are listed among the fifteen best male outdoor high jump performers for 2008. The top four jumpers for the year are white. Overall, ten of the fifteen are white. (One is Brazilian. I will leave it to others to assign him to a race.) Finally, it is worth noting that the best American jumper in 2008 was white Texas native Dusty Jonas. That must be an odd fact to swallow for all the Americans who believe white people are genetically deficient in jumping ability.[5]

Most revealing of all are the results of Olympic high jump finals. These are the big tests, the greatest international showdowns between the planet's greatest human flyers. After more than two centuries of competitions, however, only a few African Americans and one dark-skinned Cuban have ever won gold medals. Every other winner has been white. Furthermore, no black African athlete has ever won an Olympic high jump medal of any kind.[6] This is the one international competition that directly measures jumping ability and when the biggest prize of all, an Olympic gold medal, has been on the table, white athletes have outperformed black athletes almost every time. What else but race-based genetics could be behind this glaring racial disparity we see at the top levels of high jumping? Clearly there must be a white jumping gene that makes members of the white race such great levitators. We can't deny the evidence before our eyes. Right?

It is a popular belief in America and many other societies that "black people" are the world's greatest athletes, not because they care about sports more or because they work harder at it, but because they are racially gifted. Because of their race, they are born with innate advantages that make them naturally better. Why do so many people believe this? And is it true?

THE "WORLD'S GREATEST ATHLETES" ARE WHITE

The Olympic decathlon is a dramatic two-day test of various athletic abilities. Athletes compete in the following diverse collection of events: pole vault, 1500-meter run, 400-meter and 100-meter sprints, shot put, discus, 110 hurdles, javelin, long jump, and high jump. Competitors earn points in each event based on performance, and the athlete with the highest points total at the end wins. The result is a competition that does a very good job of determining who comes closest to the complete athletic package. Because the events are so different and the two-day affair so physically and mentally demanding, the winner has traditionally been given the unofficial title of the "world's greatest athlete." It seems fair to assume, therefore, that a review of the Olympic gold medalists in this unique event will reveal a great deal about racial superiority in sports.

Of the twenty-four Olympic gold medals awarded in the decathlon, it turns out that only two have been earned by black athletes (Milt Campbell and Rafer Johnson of the United States, in 1956 and 1960, respectively) and one by Native American Jim Thorpe in 1912 (it should be noted that Thorpe had two white grandparents). Many people may identify three additional gold medalists as black, but this can be challenged if the issue on the table is athletic superiority based on racial genetics. Two-time winner Daley Thompson of Great Britain (1980, 1984) has a black father and a white mother. So does 1996 winner Dan O'Brien of the United States. American Bryan Clay, the Olympic champion in 2008, is the son of a Japanese woman and an African American man. So, in the twenty-four Olympic Games that included a decathlon competition, only two saw black athletes (both parents being black) win a gold medal. White men overwhelmingly dominate the list of Olympic decathlon champions. Don't we, therefore, have to conclude that the white race is vastly superior in terms of innate athletic ability? How else could it have produced so many of the "world's greatest athletes"?

The Winter Olympics draw elite athletes from some eighty nations to compete in diverse events that test eye-hand coordination, speed, strength, and endurance. Guess who dominates? There must be something inherent in the white race—superior athletic genes—that makes white athletes naturally better at skiing, ice hockey, snowboarding, curling, the biathlon, figure skating, and so on. Oh, sure, the occasional black athlete has some success, but the overall pattern is clear. White athletes are simply better than black athletes. Results don't lie. Whites have won more than 99 percent of the Winter Olympic medals ever awarded, and they hold virtually every Winter Olympic record. The argument that blacks do not compete in the games due to racism or some other reason is simply not true. Blacks can and do compete in the Winter Olympics. For example, African American Shani Davis won gold and silver medals in speed skating in 2006, and even Kenya, the nation that produces so many great endurance athletes, competes in the Winter Olympics. In 1998, Kenyan Philip Boit competed in the ten-kilometer cross-country skiing final. He finished in last place, a full twenty minutes behind the white Norwegian winner.[7]

One cannot simply explain away the Winter Olympics as irrelevant or as the result of cultural and geographical influences that favor cold-weather societies because if we look at the Summer Olympic Games, we see more of the same. While black athletes do well in some sports, the overall picture is one of white domination as well. The reality is that most individual gold medalists in the Summer Games have been white, and the nations that have won the most medals have Olympic teams made up mostly of white and other nonblack athletes.

After the 2008 Olympic Games in Beijing, the United States leads the world's nations, with 2,514 medals won in both Summer and Winter Games. The vast majority of these medals were won by white Americans. The rest of the world's all-time, top-ten nations are the former Soviet Union, Great Britain, France, Germany, Italy, Sweden, the former East Germany, Hungary, and Finland. All of these are nations with teams that are overwhelmingly made up of white athletes. One has to go all the way down to the thirty-eighth spot before finding the first African nation on the list (Kenya, with seventy-five medals).[8]

Individually, the case for white sports supremacy is no less obvious. On the list of the top seventy-five all-time greatest Olympians (based on most medals earned, in the Summer or Winter Games), only one, track and field athlete Carl Lewis, is black. Lewis ranks twentieth with ten medals, well behind the eighteen medals of list leader Larissa Latynina, a Russian gymnast. Others on the list include seventy-two white athletes and two Japanese athletes. No black Africans make the top seventy-five.[9]

White athletes dominate international football (soccer), too. This is probably the world's most popular sport, yet no African nation has ever won the World Cup or even made it to the final match. It certainly isn't for lack of interest. I saw kids playing football virtually everywhere I went in Africa. Still, nearly every winning team in the history of the World Cup has been all or mostly white. The most notable exception has been Brazil, but most fans would probably describe that nation's victorious teams as "racially mixed" rather than mostly black. Additionally, France's 1998 World Cup–winning team had a significant number of nonwhite players.

The list of sports that white athletes excel in while black athletes struggle in them seems endless. Very few black men, for example, have succeeded in professional ice hockey. There are National Hockey League teams in many American cities with large black populations, including Detroit, New York, Chicago, Philadelphia, Los Angeles, and Atlanta. But whites rule the ice, nonetheless.

For some three decades now, the World's Strongest Man competition has been drawing behemoths from around the globe to battle against each other in creative events that test strength, power, and stamina. No black man has ever won the competition. In fact, the champions have mostly been Finnish and Icelandic athletes—about as white as one can get. In addition, black athletes have not fared well in powerlifting and Olympic weightlifting, sports that probably provide the best measure of human strength.

The world of fighting can be volatile. Champions often don't stay on top for long, but at the time I am writing this (July 2009) the world heavyweight boxing champion is a white man. Currently it is the same story in the Ultimate Fighting Championship (UFC), the world's leading mixed martial arts organization. The heavyweight champion is white. White athletes have long dominated Olympic freestyle and Greco-Roman wrestling as well.

Cycling is a popular international sport that offers fame and fortune to anyone who can pedal his way to the top. It includes a broad spectrum of events that test speed, power, and endurance. This means the sport offers opportunities for success to a wide range of athletic types. Still, white cyclists dominate the sport, from short track events all the way to the long and grueling Tour de France.

This has not been anywhere near a complete list of white-dominated sports. But the writing on the wall should be clear by now. Something is going on. Why are white people winning so many gold medals, championships, and trophies while black athletes bring up the rear or go missing entirely? Perhaps we should assume that there is a genetic explanation.

White athletes must have a racial advantage. There must be some inherent gift, unique to their kind, that gives them an edge in sports, right? What else could it be?

ANOTHER WAY OF LOOKING AT RACE AND SPORTS

I hope it is obvious to readers that I intentionally built a house of cards over the previous pages in order to make a point. I do not believe in the genetic superiority of white people in sports. While the figures I cited are all correct, there are much better ways to explain them than putting forward claims of white genetic superiority—just as there are better ways to explain the success of black athletes than putting forward claims of black genetic superiority. Before anyone points the finger at race to explain the sports results, we ought to consider factors that are far more likely to be behind who plays, who wins, and who loses. These factors include culture, access, economics, traditions of success, racism, and irrational race beliefs that impact the psychology of athletes and coaches. Maybe there are not many black men on the rosters of NHL teams because relatively few black kids play ice hockey. Maybe black athletes have not made a big splash in competitive swimming because large numbers of black people have not had access to pools or coaching in the past. Maybe white people and countries with mostly white populations have been cleaning up at the Olympics for the last two centuries because success at that level has everything to do with opportunities, resources, and investment in sport, something poorer nations with large nonwhite populations have not been able to afford or have chosen not to make a priority.

"Environmental and genetic explanations for racial domination in sports ability are difficult, if not impossible, to disentangle," according to Joseph L. Graves, an evolutionary biologist who has looked closely at various race-sports beliefs. "The first obstacle is that human genetic variation cannot be unambiguously partitioned into races. The second problem is that environmental influences that might impact physiological performance are not consistently associated with any particular population (however defined). Finally, any investigation of athletic performance must take into account social, cultural, and economic factors that influence who is likely to have the opportunity to achieve in a given sport at the highest levels."[10]

European cultural and intellectual historian John Hoberman is the author of *Darwin's Athletes*, an important book about race and sports. He sees serious problems stemming from the belief in the natural athletic superiority of blacks. "High-profile sports characterized by 'black dominance' continually reinforce surviving 19th-century ideas that identify blacks with their bodies and promote fantasies about their lower evolutionary development," says Hoberman. "Constant exposure to evidence of superior black athletic ability has turned blacks into 'natural' athletes in a way that is quietly stigmatizing, in part because they are accompanied by unending stories about black academic deficits."[11]

I asked Mike Barrowman, the 1992 Olympic 200-meter breaststroke champion and former world record holder, what he thinks about race and swimming:

> Who'd be crazy enough to publicly state that race plays a part in white athletes' current dominance of swimming? A touchy subject for sure! The fact is, though, we've had black Olympic swimming champions, so it would be hard to argue that any disadvantage derived from their race. My personal opinion is that much of the dominance of white swimmers today is due to cultural influence. From the youngest ages, we are repeatedly blasted with the faces of the greatest football and basketball players' faces. Many of these are black. And the cycle continues. Talented young black athletes see these superstars as role models. They see a realistic chance at moving out of a lower-income area, a realistic chance at attending a top quality school, a realistic chance at making contacts that will propel them into the top echelons of society. And, quite correctly, they see people who may have grown up just like them and made it to be the best in the world at something. Something they could achieve by using the same court or field just down the street. How many neighborhoods have swimming pools you can just walk into like that? How many have paid coaches to help these young people [become competitive swimmers]? Not enough. Not yet at least.[12]

WHY BOTHER?

Some people may ask, "So what? Why bother bringing something so trivial as sports into a topic as important as race and racism?" The reason is simply that sports do matter. Do not doubt for a moment that sports have a huge impact on the issue of race. The games we play and watch significantly influence the way people think about race. Things we see on the playing fields and hear from commentators help perpetuate popular race beliefs. I cannot count the number of times one sport or another has come up while I have been discussing the general subject of race with people. Sports are a very visible part of global culture. I have found that even people who are not sports fans tend to hold highly racialized views on sports. Simply through popular media exposure, by a sort of cultural osmosis, people who are not particularly interested in sports pick up ideas about race and sports and often incorporate them into their worldviews. Sports act as windows into humankind that often seem to confirm our prejudices and stereotypes. Incorrect conclusions about why the sports landscape looks the way it does can lead to dangerous and destructive conclusions about our fellow human beings. Because of this, our games, silly as they may be in the big picture, are a crucial part of the race discussion.

Although it is undeniable that culture, economics, opportunity, and other factors play a very prominent role in sports, maybe some "races" might do better in a particular sport because they have a higher frequency of genes that are favorable for that sport. Is this possible? "Beats me," says biological anthropologist Jonathan Marks. "I tend to prefer invoking real social conditions over imaginary genetic ones."[13]

Marks makes an important point. Why would anyone think it is a good idea to look at two athletes who come from very different worlds and try to explain the difference in their performance in terms of racial genetics? Explanations like that make no sense when the lives of athletes leading up to the moment of competition are so different. Why should we look for something as vague and frustratingly elusive as Wayne Gretzky's or Michael Jordan's biological race as the key to their success when other explanations, far more down-to-earth and more likely, are right before our eyes. Gretzky and Jordan are special individuals who, through some combination of individual genetic gifts, enriching environments, opportunity, and hard work were able to achieve greatness. Both men worked long and hard to make themselves into supreme athletes who transcended their games. Their skin color, nose dimensions, hair texture, and assigned membership in an absurdly broad category of humankind are the last things anyone should look at to find the reason they became champions. Their relatively trivial genetic kinship to hundreds of millions of other people who are only slightly more similar to them than anyone else did not place a crown upon their heads at birth and guarantee that they would be great. Only one black person has ever played basketball quite the way Michael Jordan did. Only one white person has ever played ice hockey the way Wayne Gretzky did. Why isn't that enough for us? Why does race trump individuality in the minds of so many people when it comes to sports?

I was the sports editor of a newspaper some years ago and exploited the job for all it was worth. I used my credentials to score an up-close view of many unforgettable competitions and to meet many great athletes. My heart raced almost as fast as Michael Johnson's golden shoes when he streaked by me on the way to a then–world record run of 19.32 seconds in the 200-meter final at the 1996 Olympics. I felt as if I was on the track when Dan O'Brien, Donovan Bailey, and Joseph Keter cruised to gold medals in their respective races. My eyes widened during a Monday night game when NFL star Jerry Rice weaved his way free from defenders to snatch a pass in the end zone only a few feet away from me. I have interviewed a long list of great sports figures and observed a wide variety of games and athletes, from the pee-wee leagues up to the pros. Through it all, nothing I have seen, heard, or read has convinced me that biological races are the best explanation for the racial disparities we see in sports.

I saw Michael Jordan play during his prime and I met him. Seeing him in action was unforgettable. I won't go too far in praising him because he's been written about and fawned over so much that it can get tedious if not nauseating. But I will say this: #23 was a rare and beautiful expression of human grace and pure determination. The way he shot, jumped, and cared so much about winning touched virtually everyone who took a good hard look his way. But whenever I found myself watching Jordan more than the game, I didn't see a "black athlete." Jordan transcended not just the game but race too. In my mind, at least, he was fully human, not a representative of just one fraction of humanity. He represented all of it. Isn't that how it should be? When we racialize athletes, don't we dehumanize them to a degree, rob them of their individuality? Don't we cheat them out of full credit for *their* achievement? I am fully aware that it is risky to speak or write of seeing through or beyond a black man's race because some might interpret it as racism, an inability to simply accept and embrace a black man as a black man. But in the context of popular beliefs about race and sports, it is Michael Jordan's "blackness" that is in some ways a distraction from his work, talent, and drive.

It is ironic that Jordan undoubtedly contributed to the popular belief that black people are naturally or racially gifted basketball players. His unprecedented success was seen as evidence, if not final proof, that black people were born to play the game. Dig a little deeper, however, and we find that this notion contradicts everything about who Jordan really was and why he succeeded. He was cut from his high school basketball team. He was not the first pick in the 1984 NBA draft (he was third), and no one expected him to become a superstar, let alone the greatest ever. But there was something special about him, and for all the hype about jumping and dunking, it had very little to do with race. Jordan immediately earned a reputation in the NBA as an astonishingly hard worker. He practiced as hard or harder than anyone on his team, probably harder than anyone in the entire league. Chicago Bulls teammate Rod Higgins witnessed Jordan's first season in the NBA and was in awe of him: "You couldn't help but notice this guy was different from all of us who were already there with the Bulls. His practice habits were unmatched."[14]

Jordan himself says it's all about the work you put in. "Players who practice hard when no one is paying attention generally play well when everyone is watching."[15] The suggestion that Jordan's biological race explains his greatness is both wrong and insulting. Race didn't make him a great player. *His* hard work and *his* individual talent did.

HOOP GENES OR HOOP DREAMS?

Basketball has significantly influenced America's perception of race. To suggest that one race or another has cornered the market in ice-hockey genes or basketball genes, however, makes no sense. Yes, African Americans are overrepresented (relative to the US population) in the NBA at this moment in history. But so what? What does that really mean? It is by no means proof that black Americans are naturally superior basketball players. If the black race were really loaded with basketball-greatness genes that made them naturally better dribblers, shooters, and rebounders than people of other races, then no white person would ever get near a court because there are hundreds of millions of black people in the United States, Africa, the Caribbean, South America, Europe, and elsewhere who would be predisposed to be better on the court. It only takes four hundred fifty players to fill the rosters of the NBA's thirty teams. If black people really are the chosen ones of the hard court, more beholden to hoop genes than to hoop dreams, then surely coaches with jobs that depend on winning would not waste their time with hopelessly inferior nonblack players when there are so many millions of black people on the planet to draw from.

Why do we see so many nonblack players in the NBA today? Racism could be behind it, but that seems unlikely when some teams often start five black players. Yes, it is possible that some racist conspiracy is at work to whiten up the league so that white customers will keep buying those tickets and jerseys. But this is pretty difficult to believe at a time in America when black athletes are so celebrated and embraced by most white people. An awful lot of white people bought Air Jordans in the 1980s and 1990s. And white fans seem to have no hesitation when it comes to idolizing LeBron James and Kobe Bryant today. The men's basketball team that represented the United States in the 2008 Olympic Games did not include any white faces. The possible exception, I suppose, could be Jason Kidd, who has one white parent, but he is identified as black by most fans. There does not seem to have been any concern with image or with soothing racists when that high-profile team was selected.

If black players are naturally better at basketball, then why do most black players in division one colleges fail to make it to the NBA every year while a significant number of white college and white foreign players do? When I grew up in the 1970s and 1980s, I constantly heard this idea of blacks being "naturally" better at basketball than whites get presented as though it were an irrefutable fact. But maybe it is only white Americans who can't play, because non-American white guys seem to be doing very well in the NBA these days. Is their whiteness somehow different from American whiteness? If white Americans can't play basketball very well because of their genes, then it is odd that those who may be seen as their closest genetic kin in the world can. Today, white and other nonblack players aren't just making it to the pros; some of them are superstars. No one doubts the abilities of Steve Nash, Yao Ming, and Dirk Nowitzki, for example.

And what about Africa, the presumed source of the prized black "basketball genes"? If African heredity is supposed to be somehow responsible for basketball success, then why have there been so few African players in the NBA? The game is popular in many African nations, and US college and NBA scouts do keep an eye on talent in Africa. But to date, only three African players—Hakeem Olajuwon (Nigeria), Dikembe Mutombo (Democratic Republic of the Congo), and to a lesser degree, Manute Bol (Sudan)—have made a big impact in the NBA. Something doesn't add up. It's almost as if there is some mysterious force other than biological race that is influencing those we end up seeing in the NBA.

THE FORCE OF CULTURE

It is very interesting that so few people try to explain, at least openly, white domination of sports such as cycling and the decathlon in terms of genetic racial superiority. Why is this? My hunch is that white athletes benefit from a subconscious racism that creeps into our thinking about sports. White athletes who excel are typically always seen as one of two things or a combination of the two. Their success is likely to be seen as the product of hard work, dedication, and sacrifice. Or they are seen as *individuals* with rare gifts. By contrast, a black athlete always seems to be seen as great first and foremost because of his or her blackness. This is not only inconsistent thinking; it's also unfair to black athletes who work and sacrifice to become good at their sports. Blindly pointing to race while ignoring all the non-genetic, social factors we can see right in front of our eyes is silly. It's just like crediting magical brain genes for the academic success of some Asians while ignoring the fact that they typically study harder and longer than everyone else because of cultural and family influences. Think about it: if black kids are naturally better at basketball, then why do they have to practice so much? Why do so many of them work night and day on improving their crossover dribble, their jump shot, and their reverse dunk? If it all comes naturally, it seems they are wasting an awful lot of time and energy unnecessarily.

The force of culture is too important to ignore. We can't make claims about biological race advantages as explaining who wins and who loses, or even who plays the game, when we know full well that athletes are not arriving at the stadium with the same life experiences. If today the majority of white kids in America fell in love with basketball, were told over and over that they could make it in the NBA, and then worked on their skills every day, it's very likely that the NBA would be dominated by white men within twenty years or so. Why wouldn't it? After all, we know that white men can jump, because they have the Olympic high jump gold medals to prove it. And we know white guys can play basketball, having seen the recent success of so many white non-American players in the NBA, in addition to the many white American players who have excelled. When we strip away all the irrational beliefs and assumptions about racial biology, culture is the only explanation left that makes sense.

"Basketball ability, just like any other human behavior, is determined by a complex interplay between individual genetic ability, personality, culture, and society," according to Dr. Graves:

> African Americans, who currently dominate the game, represent a genetically and culturally unique population, one that is not equivalent to any particular Western African population, either in genes or in culture. Success at the modern game of basketball is facilitated by speed, endurance, agility, strength, height, hand-eye coordination, and leaping ability, among other athletic traits. There is no reason to suppose that these traits are found disproportionately among people of African descent in the United States, nor is there any scientific way of separating the genetic, environmental, or cultural effects that determine athletic predisposition. Thus, any claims of African genes providing superior athletic performance are at best speculation, and at worse racist ideology.
>
> When Michael Jordan retired from professional basketball, he was asked once again by reporters why he thought black players dominated the sport. "Okay, I'll tell you," he said. As reporters leaned forward, pencils poised, he whispered into the microphone, "We practice."[16]

To build my bogus case for white racial sports supremacy at the beginning of this chapter, I only had to pick and choose what to emphasize in order to support my claims about race and sports. I cited some sports and ignored others. I played down the cultural, historical, and geographical factors that surely determine which individuals and which nations tend to do better in specific sports. I never mentioned economics, the impact of race beliefs on children who play sports, or the role of race belief and racism in influencing the sports landscape. Hopefully this shows how easy it is to be misled by a biased presentation. The success of some members of one race in a few select sports should never be seen as proof of some broad genetic superiority. At present, athletes who are identified as white dominate cross-country skiing, and athletes who are identified as black dominate long-distance running. Both are tests of stamina, muscular endurance, and cardiovascular fitness. What, then, do we conclude? The only way anyone could ever make an intelligent comment about one or the other being superior long-distance athletes is by having all white and black people grow up in nearly identical cultures with nearly identical diets, environmental conditions, opportunities, and limitations. Then, after at least a hundred years or so of perfectly controlled conditions, we might be able to determine some sort of group genetic differences. Maybe.

When Americans or any others glance at the NFL and NBA today and, based on what they see, conclude that black people are superior athletes, they are doing precisely what I did at the beginning of this chapter. That makes as much sense as looking at all the white faces of the winners of professional skateboard competitions or swimming competitions and concluding that white people are superior athletes.

Maybe the majority of players in the NFL today are black because black kids want to play in the NFL more and work harder for it than any other group of Americans. Maybe white people win so many medals in skiing because Finland and Norway tend to get more snowfall than Nigeria and Kenya. Maybe no black man has ever won a World's Strongest Man Competition because the strongest black men in America are busy toiling in the trenches of the NFL. Maybe black men have not done as well as whites in the Olympic high jump because they don't think it's cool, it doesn't pay enough, or it's not as much fun as basketball. Maybe so many white guys from Europe and elsewhere are making it in the NBA these days because, unlike some white American males, they didn't grow up constantly hearing that they were racially inferior athletes. Maybe we haven't seen a string of great black cyclists yet because young black men wrongly believe they are not suited for that sport and never try. Once we eliminate all these possibilities—and a million more like them—then we can justify speculating about biological race as an explanation of racial disparities in sports.

WHO SAYS BLACK PEOPLE ARE GREAT RUNNERS?

Here's a question to ponder. If, as many people believe, the "black race" is the best at running, then why aren't there any great East African sprinters? Furthermore, where are the great West African and African American long-distance runners? Aren't all these people members of the "black race" too? Why do so many people claim that black people are great runners? Haven't they ever watched the Olympics on television? The sprint finals are usually filled with athletes drawn from a relatively small group of people who have mostly western or central African ancestry. (Most African Americans and Caribbean blacks have western or central African ancestry.) Meanwhile, the long-distance events are usually won these days by athletes from East Africa. If black people are the fastest runners, then why don't we ever see Kenyans or Ethiopians in the Olympic 100-meter final? Are they less black than West Africans and African Americans? And how long is it going to take before we finally see an African American or a Nigerian win gold in the 10,000 meters? Why do we see more white Europeans in elite long-distance races than West Africans and African Americans? I thought "black people" were supposed to be the best runners. West Africans and African Americans are black, aren't they? Something weird is going on.

The problem, of course, is that biological race categories don't make sense. Apart from dark skin and a current or relatively recent connection to a very big continent, what do West Africans, East Africans, Caribbean blacks, and African Americans have in common? Not as much as most people assume, it turns out. The African population is very genetically diverse, more diverse than the rest of the world's populations put together. This is because humans first evolved in Africa and have been there far longer than anywhere else. So when sports fans watch the Olympics and see an Ethiopian win the marathon and a Jamaican win the 100 meters, they may think they just saw two "black" runners win, but that is a gross oversimplification of the true relationship of the runners. What the fans really saw were two people who were nowhere near as closely related as crude racial identification suggests. For example, a typical Ethiopian could be more closely related to some white Europeans or Arabs than she or he is to a typical black Jamaican. Thinking of all black Africans, as well as all black Caribbean people and all African Americans, as members of something called the "black race" is fine if you choose to do that, but keep in mind that this is a culturally created group with very little biological relevance at its foundation.

WHY ARE WE EVEN TALKING ABOUT RACE AND SPORTS?

Perhaps the biggest problem with the whole race and sports connection is the recurring, annoying problem of what race is and what it is not. In the minds of most people, race is primarily about biology, but it's not. Remember, a human being today can literally change races simply by boarding a plane and flying to another society that has different race rules. Races are mostly cultural creations, and there is no consistent definition or sensible set of rules that applies all the time in every place. It makes no sense to say that African Americans are *inherently* better basketball players, when the category of African Americans is so biologically diverse and illogically structured. In America, even a small portion of black ancestry can mean one is identified as a member of the African American race. I suppose this is fine as a cultural practice, if that's what people want. Cultures are flexible, and the rules are whatever the rule makers want them to be. But when it comes to biology, something that is supposed to be objective and scientific, this is absurd. Many millions of people around the world saw Daley Thompson, Dan O'Brien, and Bryan Clay win their Olympic decathlon gold medals, and many of these fans undoubtedly filed these events away in their minds as further confirmation of black athletic superiority. But this ignores the fact that each of these men has a nonblack parent. Many people look at the NFL and NBA and see many "black" men, but is that really what they are seeing? Besides the obvious cases of those who have one white parent, the percentage of "white" or European genes in African Americans is significant, ranging from 6 percent to 40 percent.[17] This is the recurring problem of seeing a cultural group but interpreting it as some sort of biologically pure category. Once you start from there, your ideas and conclusions are always going to be on shaky ground.

"If humans don't have biological races, you can't explain the success of individuals and groups in certain sports on the basis of a racial differentiation of genes," says Dr. Graves:

A very good example of the nonsense that passes for racial belief in sports occurred … at the 2008 Beijing Olympic Games. By anyone's judgment, Michael Phelps was clearly the most outstanding athlete at these games [eight gold medals]. Throughout his historic performances, the narrative focused on his coaching, his mother's dedication to him as a child, his single-minded pursuit of Mark Spitz, and to a lesser degree his physical attributes such as arm length and feet size. Michael Phelps has no physical features that suggest African descent. Another outstanding performer at these games was Usain Bolt, who set world records in both the 100- and 200-meter dashes. Bolt is of African descent and from the nation of Jamaica. As soon as Bolt set these records, articles began appearing in the press describing a particular genetic variant that is very high frequency amongst Jamaicans, but lower frequency amongst Americans. This genetic difference was suggested as the rationale for Bolt's and the rest of the Jamaican sprint success. One can immediately see the logical contradictions here. First, the same muscle groups are involved in sprinting and swimming. So, why wasn't Phelps' or the other European-derived swimmers' dominance in the pool the result of specialized genes for swimming? Why wasn't Bolt's success on the track the result of his coaching or mother's dedication?[18]

Graves rejects the idea that "blacks" are naturally better athletes for many reasons. First and foremost, of course, he believes that the so-called "black race" is not a valid genetic subset of humankind. However, he freely admits the obvious, that our biological diversity can have some influence on sports:

What we do know about human physical variation suggests that some populations are predisposed for success in certain sports. We don't expect Biaka Pygmies [Africans, Congo] to excel at basketball or volleyball, but we would expect the Watusi [central Africa] to excel in those sports given the opportunity. Populations that live in high-altitude regions or countries accumulate genetic traits that are beneficial to long-distance running. The Kenyans of the Kalenjin [high-altitude] region have produced a disproportionate number of long-distance runners, but the Turkana [Kenya, high-altitude] region has not, and the low-altitude Kenyans have not. Kenya isn't the only high-altitude country, of course, and by my estimate the greatest long-distance runner of all time was the Finn Lasse Viren.[19]

Dr. Graves is saying that, while genes matter greatly in elite sports and small group advantages can influence results, the popular race model of black versus white versus Asian versus Hispanic, and so on, does not hold up when subjected to scrutiny. Dr. Graves also addresses the popular idea that the success of West Africans and those of relatively recent West African descent in the sprints in recent years is the result of a genetically preordained abundance of fast-twitch muscle fibers. (Muscles include two types of fibers: fast-twitch for speed and power and slow-twitch for endurance.)

"Empirical studies suggest that the legs of a world-class sprinter would have about 80 percent fast-twitch fibers and about 20 percent slow-twitch, while the average active person would be expected to have about 50 percent of each," explains Graves:

As of 2007, there have been no systematic analyses of muscle-fiber-type distributions in untrained persons around the world.

However, much has been made of the few studies that have examined differences between "racial" groups. For example, a 1986 study examined fast- and slow-twitch muscle types between West African "blacks" and French-Canadian "whites." The authors found that the blacks were 67.5 percent fast muscle, but whites were only 59 percent. Using these averages, and applying a normal curve based on the variability in the data, the researchers concluded that the black curve would have a greater probability of producing fast-twitch percentages consistent with what would be expected in world-class sprinters. [They estimated world-class sprinters should have more than 90 percent fast-twitch fibers.] Because they believed that the differences in muscle-fiber proportion were genetically determined and therefore immutable, they claimed blacks were naturally more likely to produce world-class sprinters rather than long-distance runners. However, without a worldwide large-scale sampling of untrained individuals, there is no real way to interpret any differences in muscle-fiber composition among athletes or to make any legitimate comparisons of genetic predispositions for long-distance running. The differences may be a matter of training rather than genetic inheritance.[20]

Graves adds that easily missed factors such as diet and even conditions in the womb can influence how genes are expressed, possibly having a make-or-break impact on running and other sports abilities.[21] Again, we cannot justifiably draw conclusions about racial superiority based on the physical appearance of the people we see standing on the medals podium. There are infinite social, psychological, and environmental factors that contribute to medalists' success. Dr. Graves also brings up the question of what he believes are missing white sprinters:

> Intrinsic racial difference in muscle-fiber percentages does not explain the differences currently observed between the number of world-class sprinters of African American and European American origins. The racial typologists claim that there is a greater percentage of blacks whose range of type II muscle fibers is suitable for producing world-class sprinters. Yet even if this greater percentage were ten times more likely to produce world-class sprinters, it would still not explain the observed differences in world-class sprinter distribution. The actual number of world-class sprinters originating from a racial group would depend on the size of the population in question. Using the total population size and the relative proportions of whites and blacks in the United States, one would expect to find 303,118 blacks and 206,672 whites with the genetic architecture required to be world-class sprinters. Proportionately, there should be only 1.46 times more blacks than whites with the proper genetics to be world-class sprinters. Yet African Americans have dominated sprinting in America in the late-twentieth and early-twenty-first century in numbers greater than predicted by this theoretical distribution of genotypes. It must be remembered that this scenario assumes that blacks are ten times more likely than whites to have the proper muscle distributions, which data during this stretch of time show is highly unlikely. Therefore, other reasons must be found to explain why whites have not excelled in sprinting.[22]

Dr. Graves is on to something here. Where are the white sprinters in the twenty-first century? Even if they aren't winning, they should at least be competitive. After all, plenty of them were around in the twentieth century. Greater opportunities for white sprinters and the obstacles of racism facing black sprinters can explain some of the success of white sprinters in the twentieth century but not all of it. Many white Olympic 100-meter champions and finalists were undeniably fast, not because they beat black men but because the clock proved it. Bobby Morrow, Armin Hary, Dave Sime, and others recorded respectable sprint times on inferior tracks with out-dated training methods. There is no biological reason why there should not be a significant number of white sprinters running at or below 10.00 seconds for 100 meters today. Such an improvement would be expected, given improved training techniques, better nutrition, faster tracks, and the fact that there is more money up for grabs. (Not to mention better performance-enhancing drugs, which have been responsible for a significant number of the sub-10.0-second times that black sprinters have run.) It's not as if "white" sprint genes could have vanished in the last fifty years. Most likely, white people just don't pursue sprinting in numbers comparable to those of thirty or more years ago. They probably have other things that interest them. Maybe they are all skateboarding, competing on BMX bikes, or snowboarding. Maybe they steer clear of sprinting because of "stereotype fear." Perhaps growing up in a culture that constantly tells them they can't outrun black people negatively influences their attraction to track and field.

NECTAR OF THE GODS

The hormone testosterone has a big impact on sports performance. It is not the only factor and probably not even the most important factor in most sports, but it can help, which explains why so many athletes in so many sports keep getting into trouble for sneaking synthetic testosterone into their bodies. One of the common beliefs about race and sports is that black males naturally have more testosterone than white, Hispanic, and Asian men. This, many race believers say, explains why "black people" are so good at sprinting and make such great football players and boxers. Dr. Graves disagrees and cites a study of US military men that seemed to confirm this popular belief at first glance but ultimately did not:

> In the early 1990s, a study of racial hormone levels examined males discharged from the US Army between 1965 and 1971. The study examined more than 4,000 non-Hispanic whites, blacks, Hispanics, Asian and Pacific Islanders, and Native Americans in their late thirties. It was found that the amount of testosterone in the blood was greatest in Asian and Pacific Islanders, followed by blacks, whites, Hispanics, and then Native Americans. After the samples were adjusted for both the age and the

weight of the individuals in the groups, the order of the groups changed to blacks, Asians and Pacific Islanders, Native Americans, whites, and then Hispanics. There were a number of problems with how these data were obtained, but even if one accepts these results, they do not match the predictions of those who claim that socially defined race determines athletic performance. For example, the study predicts that Asian and American Indian men should be more aggressive than white or Hispanic men. Yet American society sees Asian Americans as the model minority group and not as likely to be superstars in aggressive and violent sports such as football or boxing.[23]

Furthermore, Dr. Graves explains that "testosterone levels are strongly influenced by daily and seasonal rhythms" and that there are many studies showing environmental conditions as having a strong influence on hormone levels. Some stressful situations can lower hormone levels, for example. "Even more revealing," Dr. Graves continues, "is that studies of hormone levels have not found any difference between the testosterone levels of African American and European American men. One study found that male testosterone level was correlated with age, body mass index [BMI], and waist circumference. When black and white males were compared for testosterone level with only age and BMI being controlled for, black males had about a 3 percent higher testosterone level. However, once waist circumference was included in this analysis, there was no difference between the groups." (The 1992 study of men discharged from the army did not control for BMI or waist circumference.)[24]

The most important point to keep in mind when one hears about testosterone levels and race is that no one can know for sure if reported differences are the result of nature or nurture. If an American black man has relatively high levels of testosterone, we cannot know if this is because he was born genetically as a black man or because he has lived his life as a black man in America.

"Once again, the same underlying problem exists in all the biological comparisons of socially constructed races: Unless all the subjects in these testosterone measurements experienced the same environments and the same social conditions, and displayed the same psychological responses to them, such studies are literally meaningless," adds Dr. Graves. "American society certainly does not treat African American and European American males equally, so measurements of hormone levels in these groups cannot be correlated to any supposed genetic differences between them, nor can it be posited that hormonal levels determine success in any specific sport."[25]

Physical anthropologist Curtis Wienker does not accept races as valid biological categories. He does, however, see the possibility that body types based on environmental adaptations may play a role in sports performance. He is not being inconsistent. Like Dr. Graves, he refers to the obvious possibility that small groups within so-called races may have some genetic advantages over other groups. This could possibly help to explain the long-distance running success of some small groups of East Africans and the absence of virtually any West Africans or African Americans at the top levels of long-distance running. So is this it? If there is a genetic component to be found that is influencing some sports results, are we reduced to pointing at incredibly small groups with favorable genes who live in a favorable environment and have some lifestyle features that seem to aid running development? And if so, so what? What does this have to do with the popular concept of race? Claiming that "blacks are naturally great sprinters" is very different from claiming that a few extremely tiny subsets of humans with favorable genes for a specific type of running may exist somewhere within a race that includes a billion or so people.

"I do not think culture can fully explain the success of some races, as you refer to them, in some athletic endeavors," explains Dr. Wienker:

> I would use the term you use, "favorable genes." Rather, if you think in terms of athletic abilities, different sports require different skills and to a degree, for example, different body shapes and proportions. A fundamental canon of mammalian biology is that all other things being equal— which they never are—equatorial environments select for linearity and proportionately longer appendages as adaptations to tropical climates, and as one proceeds toward the poles, selection pressures slowly change to favor, at higher latitudes, shorter, stockier animals with proportionately shorter appendages [known in anthropology as Bergman's and Allen's rules]. Put into very simplistic terms, an Eskimo is probably not as competitive at, for example, basketball, as for example, a Nilote [Nilotes are an African people, usually tall and long legged], all other things being equal, which they never are. It is likely fortuitous that adaptive "packages," as it were, turn out to lend themselves to success in certain athletic endeavors, which of course are cultural creations themselves.[26]

This is important. Rejecting races as meaningful biological categories and rejecting many popular beliefs about race and sports does not mean one has to overlook or pretend not to notice the fact that genetics are profoundly important in sports success and that some groups of people—small and ill defined as they may be—might well have the benefit of genes that lead to success when combined with favorable environments, cultural conditions, hard training, and opportunities. Individuals who make the Olympic 100-meter final are undeniably gifted. They were born with genes that at least gave them a shot at making it that far. Most people, no matter how hard and smartly they train, could ever make an Olympic 100-meter final. And, yes, some groups of people may even have some genes drifting about in their immediate ancestral neighborhood that could potentially aid them in specific sports-related abilities. However, a crucial point is that "groups of people" that may have favorable genes fall well short of "races" that may have favorable genes. If "black people" have vastly superior genes for running, then why are white people more successful in the sprints than Kenyans and Ethiopians? Why are white people more successful in the long-distance events than black West Africans, black Caribbean people, and African Americans? The most we can say is that some relatively small groups of people within what is commonly called the "black race" *might* have a genetic advantage over others in specific types of running. Then again, maybe it can all be explained by culture. We always have to consider that possibility, because of the very different psychological, cultural, and environmental conditions experienced by people who belong to different populations. If Kenyans in the Great Rift Valley discovered oil tomorrow and every family got a swimming pool, four televisions, and two SUVs as a result, would they still be winning Olympic medals in long-distance events fifty years from now? If the United States suffered a total economic collapse and tens of millions of little barefoot white kids had to run twenty miles to get to and from school every day, would American athletes dominate long-distance running fifty years later?

Never forget that each sport arrives at the present after countless cultural twists and turns. There is nothing logical or objective about which sports become popular and influence our perceptions of humanity. One man cares so much about the World Cup final that he suffers cardiac arrest over a missed goal. Another is so bored by it that he falls asleep during the penalty shootout. The quality of life for some sports fans rises and falls according to the fortunes of their nation's cricket team. Other fans would rather have their toenails ripped out than sit through a six-day test match. No sport is more important or more sensible than any other apart from the value we give it. It all comes down to personal taste and enculturation. For example, there is no reason why some quirk of culture and history could not have resulted in tag, dodge ball, and musical chairs being the world's most popular professional sports today. If that had happened, billions of dollars would be spent and countless hours devoted to watching, analyzing, and worshipping the best athletes in these sports. Our choice of sport shoes, drinks, and deodorants would be influenced by the reigning musical chairs champion, due to his or her various endorsement deals. And, no doubt, there would be a lot of attention paid to the question of why [fill in a race here] are *naturally* better at dodge ball than all the other races.

The truth is, there is no inherent superiority of any sport over another. We can argue endlessly about which one is a better test of human athleticism or is more entertaining but, in the end, it's a subjective call. It's opinion. We make a sport great or meaningless by giving or withholding our attention. We inflate it or relegate it to extinction according to the whims of culture. For thousands of years, our ancestors and kin have played and watched various sports with great concern and dedication. But we have lost most of these sports to time and forgotten they ever existed. They mean nothing today. Some of our modern sports, maybe all of them, will fade away as well one day. Therefore, pointing to the NFL, the NBA, or a few select events within track and field, as American fans tend to do, is hardly an objective or a reasonable basis for making judgments about something as complex and controversial as racial genetics. There is no reason to assume that the NFL and the NBA are the best measures of human athleticism. Strong ticket sales and high television ratings are an indication of current popularity, nothing more. Why not form racial beliefs based on rodeo bull riding? There can be no doubt that it demands great athleticism. Shall we then look at its top performers and judge all races accordingly? What about cross-country skiing, synchronized swimming, or bowling? Why not analyze and obsess over the racial identity of the champions in those sports and then make sweeping declarations about racial athletic superiority? If we won't do this because we suspect that participation and success in cross-country skiing, synchronized swimming, and bowling are linked to geography, economics, and cultural influences, then why do we forget that when it comes to the NFL and the NBA? What makes race believers so sure that geography, economics, and culture don't play a role there as well?

YOU CAN HAVE MY BATON WHEN YOU PRY IT FROM MY COLD DEAD FINGERS

When a skinny white kid in Florida back in the 1970s decided to give running a try, it did not necessarily mean he wanted to participate in some weird psychodrama in which every training session and every competition would be another act in the great American story of race relations. Unfortunately, the concept of race often invades our lives whether we want it there or not.

I just wanted to run. But my high school track and field career was loaded with racial baggage from day one. No matter how hard I tried to ignore it, it never really went away. Looking back, I feel fortunate to have come out the other side with so many good friends and pleasant memories. It might easily have gone another way.

I attended a fairly large public high school with a student body that had about equal numbers of blacks and whites. It was a pretty good school, not perfect, but no shootings or riots to worry about. On the first day of my first track season, I was the only white kid to turn up. Being surrounded by about 40 black kids was not alarming, but it was an indication that I might have been treading where I didn't belong. The school's cross-country team was all white, and I could cruise six miles at a good pace on my worst day. Maybe that's where I was "supposed" to be. But I was attracted to speed. I had dreams of running fast, very fast. I didn't think I was quick enough for the 100, but I felt I was just right for the 400 meters and the 800 meters. There was something about those two races that attracted me. Running close to full speed, dancing on the ragged edge of maximum effort without blowing up the engine. Relaxed and fluid. Time enough to think but no time to daydream. To me, they were the perfect races, the ideal blend of speed and toughness. To my mind, the 400 and 800 were the purest form of running, and I wanted them.

The first thing my new coach said to me that first day was: "You gonna run the two-mile, boy?" Ouch. Fortunately the two-mile race was the

longest race in standard Florida high school meets at that time or he probably would have tried to sign me up for the marathon. I felt a wave of panic
and doubt surge through me. For him to say this before ever seeing me run
was not a good sign. I wasn't dumb. I knew what was going on. "White boys
can't sprint," was the commonsense, matter-of-fact judgment on the matter.
In America, black people sprinted and white people ran long-distance
events. It was the acknowledged caste system, set in place by nature and
unalterable by anyone or anything. In south Florida in the 1970s, at least in
my town, to suggest otherwise was viewed as either crazy or some sort of
eccentric rebellion against the order of the universe.

I never forgot my feelings that day. I wasn't angry. I wasn't even sure
that my coach and my new teammates were wrong. But it didn't feel good
to be weighed, measured, and condemned, based on nothing more than a
quick glance at my skin color. Don't forget that this happened before
anyone ran one step on the first day of practice. It was tough to hold on to
what little confidence I had had in myself to begin with. Prior to that day,
I had trained alone at night, fueled by memories of the training scene in
the first *Rocky* film. I raced on foot against younger kids on bicycles. I timed
myself on an empty track and, given steady improvement, projected myself
into an Olympic final within ten years. It's a good thing for a fifteen-year-
old to have dreams. I felt fast when I ran, and I believed I could compete.
Now that belief was stumbling against obstacles and fading fast.

I learned a memorable lesson that day about the discomfort and
confidence-rattling affect that prejudice and negative assumptions can
have on a person. Trust me, no white person should ever dismiss the feelings a lone black student in an all-white classroom might have. It is not so
easy to carry on and perform well when derogatory beliefs and unfounded
assumptions about racial limitations are swirling around you.

For reasons long forgotten, I decided not to give in. I didn't bow my
head and accept the role of the team's designated white jogger, condemned
to run laps while everyone in the stands went to buy hot dogs. At the risk
of offending my coaches and, worse, being laughed at by my new teammates, I insisted that I wanted to run the 400 and 800. (It was the 440-yard
dash and 880-yard run back then, but I will refer to the races in terms of
the metric equivalents that are standard in the sport today, in order to avoid
confusion.) My dreams of becoming the next Alberto Juantorena would
not be derailed by some stereotype. I do recall thinking something along

the lines of "I am 'me,' not something called the white race." Youthful arrogance can be a great thing sometimes. The coaches, out of amusement, kindness, or pity, allowed me to take a shot at competing in those two events. So I toed the line.

I should add a brief explanation here. There is nothing wrong with the two-mile event, of course. It's a tough and beautiful race that requires tremendous dedication and talent. I have competed in long-distance events and I have no prejudice against them. Many of my sports heroes were middle-distance and long-distance runners. However, in the context of my track team back in the Dark Ages (the late 1970s in Florida), the mile and two-mile events were the land of exile, a place in which dead weight could be forgotten. It was where people who could not run went to run. This was ridiculous, of course, but it was the reality of my world.

That first season, I was set to take on the best high school quarter-milers in Florida and run through every racial stereotype that stood in my way. It would be great if I could now describe how I won every race and struck a blow against prejudice. But I can't. My team and my region were loaded with talent and the sad truth is, I sucked that first season. I was so slow that a white runner even beat me during a preseason meet in which we competed against some wimpy private school. Beaten by a white runner. Oh, the shame! That season wasn't a total disaster, I suppose. I did well enough to earn a varsity letter. But I didn't feel fast.

I could have slunk away after that mediocre year, never to wear spikes again. I could have become one more convert to the cult of racial sports stereotypes. It would have been easy to dedicate myself to the emerging computer industry and vow to buy a sports team one day and have my revenge. But no, after a little soul searching and a lot of rationalizing, I came to the conclusion that I could still do it. The only thing missing was more intense and more intelligent training. So I devoured every track book I could get my hands on. I learned everything possible about training, racing, eating, weightlifting, stretching, soaking, shoelace tying, and so on. I think I could have qualified as a division-one track and field coach by my junior year in high school. But this all seems a bit weird in hindsight, because I should just have been a kid trying to be a track star so he could impress girls. However, it seems I was more like Dr. Frankenstein trying to create life from the inanimate tissue that was my pasty white legs. Through sheer genius and willpower, I was determined to transform the body of a slow, plodding white boy into a sleek running machine that skimmed the earth and never faltered in the home stretch. My DNA might have had "nerd," "writer," or something similar written all over it at my conception, but now I would hack in and rewrite the code.

I told you it was weird. But here's the strangest part of all. It worked. I trained alone all summer, swore off all candy and junk food, and unleashed the runner within. I succeeded in turning myself into a competent and competitive athlete. In my junior year, I earned a spot on my team's 4 × 400-meter relay squad and was district champion at 800 meters. I earned the respect of my teammates, and once or twice I even heard "Not bad for a white boy," a coveted certification that I had accomplished something. I finished that season perfectly poised for some real high school glory in my senior year. However, just when I was on the brink of greatness (in my mind, at least), my senior season was wiped out by a stress fracture. Looking back, it seems clear that I overtrained in the off-season. I guess I wasn't as brilliant at self-coaching as I thought. While missing out on a season that likely would have been filled with victories was heartbreaking at the time, I have no regrets about any of it today. It was a great experience that made me tougher and more confident in later life. It allowed me to briefly taste life as a racial minority who is prejudged to be inferior. I also learned that one could breach racial boundaries without immediately bursting into flames. My track days were about so much more than running around in a circle. They forever changed the way I would view people who were in racial groups other than mine.

Being on that team gave me access to another world. It allowed me to take a close look inside the world of fellow students who happened to be black. My town had at that time near total residential segregation. Literally, there was a "White Town" and there was a "Colored Town" that everyone spoke of openly. White racists called the latter "Nigger Town." I had gone to integrated schools since first grade, so hanging out with black kids was nothing new for me. There was something very different about being the lone white boy on the track team, however. Apart from a couple of pole vaulters, and not counting the occasional two-miler who briefly came and went, I was the only consistent white member of the team. On bus rides to meets, we blasted out "Shake Your Body Down to the Ground" and "Fantastic Voyage." We cheered for each other during races, we joked, and we talked. I didn't "turn black," in my head or otherwise. But there was an obvious reduction of the distance between us that race belief had imposed upon us at birth. Through the experiences of training together and stressing over meets, we drew closer. There is something special about surviving a sadistic interval workout with friends. It strips away a lot of the cultural crap that separates people. I suppose it is something like a diluted version of combat. When you round the final curve—arms flailing and legs failing—of the tenth 400-meter run of the session, you feel tied to the poor souls in the adjacent lanes who are suffering with you. After all these years, I still remember their names, their faces, even their individual running styles. Eugene, James, Willie, Gene, Anthony, "Catman," Melvin, "Bones," Terry, Victor. I'll never forget them.

I became very close to a few of my teammates, even visiting their homes in "Colored Town." Previously this had been a mysterious and dangerous land where no white people dared to venture. Not all of my track team friends were poor, but most were. I saw their single mothers, their lifestyles that were very different from mine. I saw the extraordinary challenges of their relative poverty. My family was barely middle class but I learned that I was far better off than many other people. The best thing of all is that, somewhere along the way, my friends stopped being "black" people in my head and became simply "people"; they were kids like me. I never saw it crumble and I don't recall ever consciously chipping away at it, but the social wall between us fell away, nonetheless. At no point in my life did I make a conscious decision to see people as individuals first and foremost rather than as representatives of racial groups. It just sort of happened all by itself. But I suspect those days on my high school track team had a lot to do with it.

One thing still embarrasses me about my track days. There is one regret. I took pride in beating black runners. I loved nothing more than facing black guys from other schools, who assumed they would beat me because I was white, and blowing past them on the final turn. Sometimes competitors really did say something before races about my being white and not having a chance, but this was rare. In most cases, I could not know if they really thought it. But that never stopped me from imagining that they were thinking it. I was cool when I won races. I put on the humble act, like it was no big deal. Inside, however, I was screaming, boasting, and dancing. Every win was a double victory. I wasn't just good enough to beat them; I was so good I even beat nature. I outran the rules of race, too.

Today, fortunately, I have enough sense to see how wrong I was. I had bought into the whole ridiculous belief of biological race as destiny. While the social wall that separated black people from me fell, I never stopped believing that there was a biological wall still there. All the nonsense about racial talents and limitations had soaked deep into my mind. Why did I assume that I had to start from scratch to make myself into a runner? How did I know that I had no natural talent waiting to be uncovered? And how did I know that all of my black teammates and black competitors from other schools were loaded with natural talent? I firmly believed that I relied on smart hard work while they relied on genetic good fortune. How could I know such a thing? Why did I think this way? Weren't my black teammates right beside me during all those hard workouts? Didn't they sometimes collapse in the infield just as I did after a particularly cruel workout? They lifted weights, ran the stairs, stretched, did all the necessary hard work, and performed well as a result—just like me. A couple of my teammates were pure genetic freaks, no doubt. No one eats French fries, avoids training as much as possible, and then burns through one hundred yards in 9.5 seconds without something special in his genetic makeup. But most of my teammates, almost all of them, had to work hard to make themselves into competitive athletes—just like me. Because I believed that their racial affiliation meant they were coasting on unearned genetic gifts, I ignored and disrespected their discipline and dedication—even though I witnessed it every day at practice.

What about those black runners from other schools, the ones I beat and then gloated over, even if it was only in the privacy of my thoughts? I didn't know how hard they might have trained in the weeks, months, perhaps years leading up to that day. Any one of them might have been smarter and more dedicated to their training than I was. Maybe they had come to the starting line with less innate talent than me but compensated with more determination and a better work ethic. I had no right to take the role of the biological underdog. Because of race belief, I assumed a great deal about people I did not know. Furthermore, I disrespected my parents and every one of my ancestors—straight back to Europe and on down to Africa—by assuming that my DNA was void of athletic potential. My father was adopted as an infant, and I have no idea who my ancestors are on his side of the family. I look white, but for all I know, Jesse Owens or Eddie Tolan might be in my family tree somewhere. Remember, when it comes to race in America, anything is possible. Today I can only wonder how fast I might have been if I had believed in myself more than I believed in race.

THE POWER OF RACIAL STEREOTYPES

Although they may have little or no basis is reality, we can never dismiss the power of stereotypes and the influence they have in the real world. Racial stereotypes about sports explain why I was the only white person to show up for the first day of high school track practice and why it was initially viewed as odd that I would want to compete in anything other than long-distance events. It is unlikely that I had the greatest potential of all the white kids in my school, but I was the only one who took a shot at the 400 meters. But all the other white kids were no shows, perhaps too psyched out even to try.

Racial stereotypes do real damage, because they can unjustifiably derail dreams and make doors of opportunity appear closed when they are not. Researchers have built up a large body of work that shows how people will underperform in all kinds of tasks because of what is called "stereotype threat." Female test takers, for example, tend to perform poorly on math tests when reminded of the gender stereotype that says men are better than women at math just before they begin the test. The power of stereotypes to negatively impact sports performance is real as well. Studies have found that putting in golf, for example, can be negatively influenced simply by mentioning a negative stereotype that applies to a player, just before he or she putts.[27]

So it seems likely that the sports world, particularly in race-obsessed America, is not only shaped by dreams, genes, training opportunity, and economic resources. It is also influenced by "stereotype threat" and, perhaps, by something we might call "stereotype confidence." Something is wrong in sports when large numbers of white kids don't even try to compete in the sprints and large numbers of black kids don't think they have a chance of making it in competitive swimming. We know white people can run fast, because Olympic history is filled with fast white people who won medals. Kevin Little, a white American, won the 200 meters at the 1997 World Indoor Championships. His personal bests are 10.13 seconds in the 100 meters, 20.10 in the 200 meters, and 44.77 in the 400 meters. Roger

Black, a white Brit, took silver behind Michael Johnson in the 400 meters at the 1996 Olympics (recording a personal best of 44.37 seconds). We know black people can swim fast, because black people have won Olympic medals in swimming. Black swimmers Cullen Jones and Anthony Ervin both won relay golds in recent Olympic competitions. They didn't sink due to heavy bones or whatever; they won *gold* medals. The fact that black people have not won as many Olympic swimming medals as white people is far easier to explain by citing cultural reasons than by citing biological ones.

Jeremy Wariner, a young, white American, had the audacity to dream of winning an Olympic gold medal in what was widely perceived in America to be a "black man's event." Against all those confident assumptions about racially predetermined sprint results, Wariner won the 400-meter gold medal at the 2004 Athens Olympic Games and then added another in the 4 × 400-meter relay. The entire black race—believed by many to be the superior running race—could not produce a single man capable of outsprinting Wariner during big races in 2004. That year, Wariner won the 400 meters at the NCAAs (indoor and outdoor), the US Olympic trials, and the Athens Olympics. In 2008, however, the black race took him down a notch. One man, LaShawn Merritt, forced Wariner to settle for an Olympic silver. Still, not bad for a white boy.

To his credit, Wariner seems utterly uninterested in the race fetish that so many fans bring to the sports world. By all indications, he is content to view himself and his rivals as nothing more than people and fellow athletes. Imagine that. Veteran track and field reporter Gary Smith wrote a memorable piece about Wariner for *Sports Illustrated*. In it he described the young Texan's surprising rise to greatness as well as the variety of reactions his race gets from observers. ("Is he mixed?" a stranger asked his mother after Jeremy won a race in high school.) But it is Smith's attention to Clyde Hart, the veteran coach at Baylor University who turned Wariner into one of history's best, that stands out. Smith presents the story of Hart as follows:

> Imagine that your first day on the job was in 1957 in Little Rock, where you're assigned to search lockers for bombs while busloads of white people prevent nine black children from walking through the doors of your high school [a reference to the historic racial integration of the Arkansas school], requiring the 101st Airborne to occupy the school for the rest of the year

to prevent violence.... Imagine that six years later you're promoted to coach at Baylor, in perhaps the most prestigious athletic conference in the land—the Southwest Conference—where not a single black man competes in track and field...until you bring in one of the first two, Ronnie Allen, just before the '70s roll in. You turn around, and it's the '90s already, and you're the coach of a black millionaire sprinter named Michael Johnson as he stands in the middle of an Olympic stadium in northern Georgia and flings his golden shoes into the adoring crowd. Turn around again and you're the coach of a gold-medal-winning Southern white youth who chooses the look and style of a black youth as he sprints right through what's left of the barriers because he never even sees any barriers. Imagine all that occurring in less than 50 years, in the course of one man's working days, and then smell the air again.

"People just don't realize what sports has done in this country," says Hart. "It's mixed kids together and made them realize that some things they've been taught about people aren't true. It's been the great equalizer in America.

"There's absolutely been a barrier for white sprinters in America. There's a stigma there. White kids think that it's a black kids' sport, that blacks are superior. There are plenty of white kids with fast-twitch fibers, but they've got to get off their rumps. Too many of them would rather go fast on their computers in a fantasy world. It's not about genes, although they may play some part in it. It's about 'Do you want it badly enough?'"[28]

SEEING WHAT WE WANT TO SEE

Confirmation bias, that tendency we all have to emphasize and remember things that support our beliefs while ignoring or dismissing things that contradict our beliefs, probably plays a huge role in popular beliefs about race and sports. For example, a typical American basketball fan might see a black player pull off a surprisingly creative scoring move. He or she might then interpret the observation as more evidence in support of the claim that black people are genetically superior basketball players. "Wow, that was a great play! *Black people* really know how to play this game," the fan thinks. But then that same fan, just minutes later, sees a white player perform a play that is no less impressive. This time, however, the fan may interpret the player's move as impressive, but only for that individual.

"Wow, that was a great play! *He* really knows how to play this game," thinks the fan. Because of confirmation bias the fan does not think, "Wow, that white player is good. It must mean that members of the white race are great basketball players." No, when the observation does not match the established belief (for example, that black people are the best jumpers or best basketball players), it is psychologically finessed so it will not threaten the belief. This is how millions of American sports fans can watch Olympic track and field competitions on television, see white people win the high jump event, and still believe that white men can't jump.

Things have improved, but back in the 1980s and 1990s there was an absurd double standard hovering over the quarterback position in the NFL. When a white quarterback threw an interception it was a bad choice by the player. But when a black quarterback threw an interception it was seen by many as evidence that the position was somehow beyond the abilities of the entire black race. I am not a mind reader, of course, but I think it is reasonable to assume that this sort of inconsistent thinking goes on all the time with regard to racial sports stereotypes. White athletes who may be profoundly gifted are still more likely to be seen as "hard workers" and "smart players." Meanwhile, black athletes, no matter how hard they work and how smartly they play, are likely to be seen as products of an athletic race, the lucky winners of some racial genetic lottery.

There is a lot of money in sports these days, particularly in the United States, where the *average* annual salary of an NBA player is more than $5 million now, for example. There is also tremendous fame to be had, with great athletes rivaling presidents and pop singers in popularity. On the surface, it would seem that big-time collegiate sports and professional sports have been good for African Americans. They have boosted pride. They have opened doors. Some racist positions have been softened. Big money has flowed to some black families. But has success in sports really been so good for black people in general? Not everyone thinks so. I certainly don't think so.

Some critics view black athletic ability as something that is being exploited with very little benefit. Yes, a fortunate few get to be famous millionaires, but what about all those black kids, generation after generation, who invest everything they have in sports and then fall short of the college scholarship or the big payday? Overall, is black America really better off for its love affair with big-time sports?

A troubling truth is that, historically, white America has been far more comfortable with black people excelling in pursuits of a physical nature, such as dancing, singing, and sports, but not so comfortable when it comes to more intellectual pursuits. Yes, the modern black athlete has been cheered and embraced by millions of white fans, but I wonder how much damage the image of that supreme black athlete has done in terms of reinforcing the racist belief that black people have strong bodies and weak minds. The problem of black basketball and football college stars who fill coliseums with adoring fans on Saturdays but can't handle the classwork on Monday is well known. Some black student-athletes do very well, of course. They trade sports skills for a university degree and do better in life as a result. Sadly, however, many black athletes are not getting much out of sports in terms of long-term success and happiness. Some even pass through university campuses, playing the game and enjoying star status for a few years, without ever learning to read.[29] Experts say the problem has improved with tougher requirements for eligibility, but there is still the glaring problem of black student-athletes graduating at a lower rate than white student-athletes. A 2008 study by the Institute for Diversity and Ethics in Sport found that 76 percent of white student-athletes graduated from FBS schools (formerly called Division I-A schools) compared with 59 percent of black student-athletes at the same schools.[30] Dr. Hoberman, author of *Darwin's Athletes*, sees sports in America as a reflection of severe racial problems in the larger society. He also sees the situation in sports not as one race exclusively exploiting another but more as two races dancing in step—to a bad tune. Hoberman writes: "The entrapment of African Americans in the world of athleticism is the result of a long collaboration between blacks seeking respect and expanded opportunity and whites seeking entertainment, profit, and forms of racial reconciliation that do not challenge the fundamental assumptions about racial difference."[31]

Hoberman describes the paucity of black coaches in big-time college football as scandalous. "There does seem to be a gradual racial integration process underway in the NBA and NFL that puts blacks in positions of authority rather than simply on the field," he says. "The glaring discrepancy is in Division I college sports, where the number of black head coaches at well over one hundred schools is currently about five. This is a scandal that cannot seem to get traction, probably due to the latent racism of many white alumni who want to be able to identify with the football coach to meet certain emotional needs of their own."[32]

Hoberman draws a direct line between the centuries-old view that black Americans were best suited for manual labor rather than for more intellectual pursuits and the enthusiastic embrace of the belief that they are better at sports than they are at thinking. The physical nature of sports allows white racist fans to cheer black superstars such as LeBron James or whoever is the hot NFL running back this season while the fans can still imagine that they are mentally superior to the superstars. Hoberman writes in *Darwin's Athletes*:

> The black athletes who today refine their athletic skills and little else at American universities are thus the damaged inheritors of an educational philosophy that once promoted manual training as the highest cultural achievement to which black youngsters should aspire.[33]

> The sports fixation is a direct result of the exclusion of blacks from every cognitive elite of the past century and the resulting starvation for "race heroes"; it has always been a defensive response to the assault on black intelligence, which continues to this day. That is why the sports syndrome has made athleticism the signature achievement of black America, the reigning symbol of black "genius." Attachment to athletics has been a coping strategy for dealing with "the 'Negro inferiority' complex" identified by Gunnar Myrdal in *An American Dilemma* in 1944. This syndrome "cannot be admitted publicly," he said then, and it remains difficult to talk about and for that reason has received little systematic study.[34]

> Black America has paid a high and largely unacknowledged price for the extraordinary prominence given the black athlete rather than other black men of action, such as military pilots and astronauts, who represent modern aptitudes in ways that athletes cannot.[35]

Finally, Hoberman warns of the influence that belief in the racial supremacy of black athletes has upon children. "The sport fixation damages black children by discouraging academic achievement in favor of physical self-expression, which is widely considered a racial trait.... Educators who think about solutions to this crisis see themselves in direct competition with the sports world."[36]

Hoberman is not exaggerating. I saw this sports obsession among young black Americans firsthand while growing up and even then had a sense that there was something wrong about it. The vast majority of black males that I knew in my childhood and teen years harbored intense waking dreams of professional sports stardom. There was nowhere near such a consistent focus on sports among the nonblack kids I knew. I remember one severely obese black boy in high school who was not in the least concerned about his weight. He saw himself as a future NFL lineman, just biding his time until he was recruited by a big college and then drafted by a pro team. He seemed to view schoolwork as nothing more than an annoyance until the NFL called. The call never came. Most boys of all races dream of sports stardom at one time or another. I certainly did. But there was something much more widespread and deeper about the sports ambitions of the black boys I knew. Whereas I might speak of "wanting" or "hoping" to make it; they "were *going* to make it" or "*had* to make it."

I have no idea if this was coming from the "I can escape poverty through sports" idea or from something more disturbing, such as, "I'm black, therefore sports is all I can do." Many nonblack kids may feel overshadowed and excluded from sports due to racial stereotypes, but, sadly, I suspect that many young black boys feel trapped by racial stereotypes. Imagine being held hostage by a belief that says you are only good at running and jumping. And then imagine how you would feel when you aren't good enough to get a college sports scholarship or make it to the pros. While white kids may be discouraged or even hindered by race beliefs in some school sports, I think too many black kids likely suffer something far worse. They are not encouraged or they are outright discouraged from much of what is an infinite variety of nonsports opportunities. Sports are fine, but there is a lot more to the world than layups and touchdowns. Something is tragically wrong when children are led to the belief that they are "supposed" to be great athletes because that is one of the few things that "their kind" is any good at. I asked Hoberman what he would do if he were the father of a young black boy in America today. Would he encourage his son to work hard toward a professional career in football or basketball? He said no, unless he found that it was the only way he could motivate the boy to stay in school. But, given the astronomical odds against making it in professional sports, he would do everything he could to steer his son toward another path.[37]

By the way, I knew many outstanding black athletes in elementary, middle, and high school. Two made it to the NFL. None made it to the Olympics or the NBA.

WHY HAS THERE NEVER BEEN
AN AFRICAN 100-METER CHAMPION?

An interesting fact that might have escaped the notice of those who believe that black African genes are the fastest genes of all is that no black African has ever won the Olympic 100 or 200 meters. Every black Olympic 100-meter champion to date has been American, British, Canadian, or Caribbean. Yes, these black athletes have recent African ancestry, of course, but if African genes are the key to world-dominating speed, then who but Africans should be winning gold medals in the Olympic sprints? If it's all about the "African race," then Africans who are descended from Africans, never left Africa, and have no extensive history of mating with white people should be dominating the Olympic 100 meters. Strange, isn't it?

I have interviewed many great black athletes, up close and in person, including Pele, George Weah, Edwin Moses, Lennox Lewis, Evander Holyfield, Evelyn Ashford, Marion Jones, former 100-meter world record holder Tim Montgomery, former 100-meter world record holder and Olympic champion Maurice Greene, West Indies cricket superstars Garfield Sobers and Viv Richards, and many others. It is my pleasure to report that they are all very human. Yes, some of them oozed confidence and carried themselves with a noticeable physical grace, but not so much so that I ever sensed I was in the presence of a member of a superior subspecies of humankind. I wonder if some nonblack people end up with hardened stereotypes in their heads because they do not interact with many black people in their daily lives. Because anyone who spends some time around a good mix of black people—whether in America, the Caribbean, or Africa—has to be delusional or blind to believe that the entire black race is athletically superior. Just like any other race, the black race has a large number of people who are about as far from athletic as a human can get. One might argue, of course, that while tens of millions of black people are not fast runners or good basketball players, the genetic potential for these abilities was there when they were born. But, of course, one could say the same thing about any other race. To think of an elite selection of black champions as evidence of overall black athletic superiority makes no more sense than pointing to a bunch of white Nobel Prize winners and concluding that all white people are smart.

It should be clear that the "black race" is not the greatest running race.

It can't be, because white Europeans dominate East Africans in sprinting and white Europeans dominate West African, central African, black Caribbean, and African American athletes in long-distance events. (The African American category as defined here excludes recent immigrants to America from East and North Africa, such as Moroccan-born Khalid Khannouchi, who is currently America's best marathoner, and Kenyan-born Bernard Legat, a middle-distance star.) The most one can say at this point is that *some* West Africans (and their relatively close kin in North America, the Caribbean, and Europe) *might* have a genetic edge on the rest of the world in sprinting. *Some* East Africans and North Africans *might* have a genetic edge on the rest of the world in long-distance running. But this is highly questionable, because it is based on today's results, as if nothing before counts and as if we know what the future holds. Who can say, for example, that a surge of runners from the Andes highlands in South America won't rise up to dominate long-distance running thirty years from now? If that happens, of course, many people will say that it is "obvious" that indigenous Andean people are the athletically superior race.

Back in the 1980s, the middle distance events were dominated by white British runners Steve Ovett, Steve Cram, Peter Elliott, and Sebastian Coe. They were kings at 800 and 1500 meters. Ovett won the 1980 Olympic gold at 800 meters and Coe won the 1500 meters in 1980 and 1984. Along with Coe and Ovett, Cram and Elliot repeatedly ranked in the top ten of the Track and Field News World Rankings for the 1500 meters. Their domination was undeniable. From 1977 through 1988, a British runner topped the world rankings in every year except two (1984 and 1987, when the Moroccan Said Aquita topped the rankings). Was their excellence race based? Was this streak of success enough to declare white people the superior race of middle-distance runners? Today it's East Africans and North Africans. Nobody knows who it will be tomorrow. Given enough time, things change. They always do.

INTO THE GREAT RIFT VALLEY

During a visit to Kenya, I was eager to make it to a very special place. As a former competitive runner with a lifelong interest in the sport, I made sure to visit the Great Rift Valley. It was a double treat for me, because for anthropology addicts it's something close to sacred ground. Some of history's great hominid fossil discoveries have been made there.

I was not disappointed. The thin cool air, the red clay earth, the deep-green bush, and the constant rise and fall of hills make this place a dreamy vision of a runner's paradise. Of course, rural Kenya is not a paradise for anyone who cares to look beyond the porch of the safari lodge. There are not enough good schools and hospitals, for example, and life is hard for most people. Still, I could not help but be swept up in the romanticism of the Great Rift Valley. I recall standing on a roadside, mesmerized by the sight of a man training. He was a magnificent runner. He had perfect posture, seemingly effortless movements, and he was deceptively fast. Based on his form and pace, I assumed he must be some Olympian with a few medals on his resumé. But here, in the land of runners, who knows? He might have been the worst athlete in the neighborhood, just out for a jog in an effort to keep his weight down, or maybe he was just hurrying home after a long day at work. No way to know here.

I had a discussion about Kenya's great running tradition with two Kenyan men from Nairobi. While both were proud of their country's success, it was immediately apparent that they spoke and thought of elite Kenyan runners not as "us" but as "them." My two friends were as much in awe of the great runners of this region as I was. It was as if they were talking about a race of humans that were very different and far removed from themselves. (An interesting side note: I asked one of the men, a Kikuyu named Stephen, if he felt any connection, kinship, or bond with black Americans. "Not at all," he replied. "They have nothing to do with our people. They originally come from other parts of Africa.")

Most Westerners are probably not aware of it, but the vast majority of Kenya's elite runners belong to the Nandi and a few other tribes that are collectively known as the Kalenjin. Despite being a small minority in the Kenyan population, Kalenjin runners have won the majority of Kenya's Olympic medals. This means that describing and discussing great runners from Kenya as simply "Kenyans" is incomplete and misleading. Though it is not as misleading as thinking of great runners from Kenya as simply Africans or members of the black race. Imagine, for example, if every American professional basketball player were to be born and raised in a few small towns in North Dakota. Would it make sense to ignore that detail and refer only to "American professional basketball players"? Wouldn't that be misleading people about the real source of the best players?

For those who are determined to use race to explain current trends in competitive running, the best available option might be to claim that there is such a thing as the "West African race" and such a thing as the "Kalenjin race." Then maybe you could say something like "the 'West African race' currently dominates sprinting" or "the 'Kalenjin race' is a dominant force in distance running today." Be warned, however, that you are turning on a faucet that won't be easy to turn off. You might also have to accept the existence of something called the "white Bulgarian race" of Olympic weightlifters, the "white southern country boy race" of stock car drivers, the "white Californian race" of beach volleyball players, and so on.

Ethiopian runners also have a great tradition of running excellence. The barefoot marathoner Abebe Bikila achieved an extraordinarily graceful victory at the Rome Olympics in 1960. The more recent sight of Haile Gebrselassie, a relentless running machine if ever there was one, churning his way to gold in the 10-kilometer race is the stuff of legend. But, just like running success in Kenya, Ethiopian running success is a highly localized phenomenon. A 2003 article in the journal *Medicine and Science in Sports and Exercise* reported as follows about world class Ethiopian runners: "Elite endurance athletes are of a distinct environmental background in terms of geographical distribution, ethnicity, and also having generally traveled farther to school, often by running. These findings may reflect both environmental and genetic influences on athletic success in Ethiopian endurance athletes."[38] The article suggests that if there is a genetic component to Ethiopian distance running success, it appears to be extremely limited in scope, affecting only a tiny subset of the nation's population. This means we are again left with nothing better to say than that *some* Ethiopians may be genetically superior distance runners but most Ethiopians are not. So, if people are pointing to great Ethiopian or Kenyan runners as evidence of some genetic supremacy in long-distance running, they really can't accurately claim that it applies to black people in general, East Africans, or even specifically Kenyans and Ethiopians.

Finally, consider the following excerpts from a 2006 research article titled "Genetics and the Success of East African Distance Runners":

- Non-genetic explanations for the success of East African runners in international athletics include the suggestion that they enjoy a psychological advantage mediated through stereotype threat, or that the distances East Africans ran to school as children served them well for subsequent athletic success. It has recently been shown that elite East African distance runners traveled long distances to school as children and mostly by running; many traveling phenomenal distances such as upwards of twenty kilometers each day. One study showed that East African children who had used running as a means of transport to school had a maximum oxygen uptake (VO2 max) some 30 percent higher than those who had not, therefore implicating distance traveled to school as one of the determinants of East African running success.

- So, even within Ethiopia and Kenya, regional disparities in the production of long-distance runners have been shown. In a study of the demographic characteristics of elite Ethiopian athletes, 38 percent of the elite marathon runners were from the region of Arsi, which accounts for less than 5 percent of the Ethiopian population. These findings were mirrored in Kenya, where 81 percent of the best Kenyan runners originated from the Rift Valley province, which accounts for less than a quarter of the Kenyan population. Although some believe that this geographical disparity is mediated by an underlying genetic phenomenon, it is worth considering that these regions are altitudinous, even relative to the rest of East Africa, and that endurance athletes have long used altitude training to enhance endurance performance.

- Despite the speculation that African athletes have a genetic advantage, there is no genetic evidence to date to suggest that this is the case, although research is at an early stage. The only available genetic studies of African athletes do not find that these athletes possess a unique genetic makeup; rather they serve to highlight the high degree of genetic diversity in East Africa and also among elite East African athletes. Although genetic contributions to the phenomenal success of East Africans in distance running cannot be excluded, results to date predominantly implicate environmental factors.

- Another study found that skeletal muscle fiber type distribution did not differ in elite Kenyan runners (senior and junior) from their Scandinavian counterparts, although the senior Kenyan runners had a tendency for higher muscle capillarity than the Scandinavian runners. However, the Kenyan junior runners had lower muscle capillarity than both the Scandinavian and Kenyan seniors, suggesting that training and not genetic endowment was more likely to be responsible for the tendency of senior Kenyan runners to have higher muscle capillarity. These initial studies of elite Kenyan athletes demonstrate that factors such as increased childhood physical activity and hard training are probably the major contributors to the superior performances of Kenyan runners.

- As many of the top Kenyan runners are from rural areas rather than towns and cities, a recent study investigated the extent to which the VO2 max training response differed between Nandi town and village boys. This study was designed to investigate if the trainability of VO2 max, which has been shown to be genetically influenced, was greater in the Nandi boys from rural areas compared to those from town. No difference was found between the increase in VO2 max of town and village boys, while the magnitude of the training response was similar to that previously reported in white boys. These findings may serve to demystify the success of the Nandi runners, showing that they may not have an obvious genetic advantage in aerobic capacity or its trainability.

- An additional finding that opposes genetics as the primary factor for the geographical imbalance in athlete production was that the rural boys were more physically active than the urban boys. Clearly, increased childhood physical activity levels are strongly implicated in the success of East African runners in international competition.[39]

Anyone who wants to think of relatively small tribes and specific population groups as races can do so, I suppose. (After all, there are no agreed-upon rules when it comes to making up races.) But once we head down that path, where does it end? How far can we go in creating racial groups and hierarchies by drawing imaginary borders around groups of people who do

well in a given sport at a given point of time? If we do it for small groups
that produce successful distance runners (at this moment in time), then
why not do the same for darts (currently dominated by white Englishmen),
or the hammer throw (currently dominated by white Eastern Europeans)?
Are southern white men a genetically superior race when it comes to driving a car around an oval at high speeds? Based on recent NASCAR
results, a race believer might have to consider it, I suppose. White
Canadians are very good at curling. If they don't win, they always seem at
least to be among the medals at the Olympic Games. Does this mean white
Canadians are the superior "curling race," with far more genetic riches
with regard to that sport than white Americans in Michigan? Of course, we
could always try to go down a very different path. We could listen to the
anthropologists and accept the reality of our biological diversity. That is,
we could accept that culture matters and that we are a species with diverse
abilities spread around in ways that are not convenient or appropriate for
dividing us up into traditional race groups.

While race, genetics, and performance-enhancing drugs get more
attention these days, we should not overlook the mind. It may well prove
to be the final frontier of sports performance. A few champion athletes
have told me that there is not much difference, physically, between athletes at the highest levels of sport. What sorts out the winners and losers,
they say, is the mind. Who wants it the most? Who can stay focused in
training during the long, lonely off-season? Who can shut out the noise and
deliver her or his best performance on judgment day? An interesting article
written by researcher Bruce Hamilton and published in the *British Journal
of Sports Medicine* raised the possibility that psychology may be a key factor
behind the success of some Kenyan and Ethiopian runners:

> Domination of individual sports by countries or regions of the world is
> not a new phenomenon. It seems that the presumed causes of such dom-
> ination are often recycled, out of date, and based on misinformation and
> myth. Over the last few years, it appears that North African countries
> have been producing large numbers of elite international athletes. Are
> we now going to search for the genetic advantages of these nations?
> Although there is no conclusive evidence for an inherited physiological
> advantage to the East African, this does not exclude the possibility that
> one actually does exist. It may be that the technology required to detect
> any differences is currently lacking.

However, irrespective of the existence or otherwise of any physio-logical advantage, it is possible that the attribution of Caucasian running "failures" to anecdotal stable external factors disempowers Caucasians. Similarly, this attribution style empowers the East African, just as it did the Scandinavians in the early 20th century and Australasians in the 1950s and 1960s, with a psychological advantage, the importance of which cannot be overestimated. Fixed beliefs and attitudes to those anec-dotal contributory factors continue to impede the success of Caucasian athletes. Sports scientists and practitioners aim to maximize athletic per-formance, and yet it seems that they are all too ready to accept that the East African dominance is due to factors out of their control. Although many factors contribute to East African running success, present Cau-casian belief and attitude systems may be a significant perpetuating influence. Until our athletes, coaches, and support staff accept responsi-bility for their own performance, the current level of athletic domination by East African athletes may continue.[40]

In appreciation of his enlightening work on race and sports, I'll let evolutionary biologist Joseph L. Graves have the last word: "If you tell a Euro-American kid he can't play basketball because he cannot jump or an African American kid that he can't swim because he can't float, you limit what they can become and further reinforce segregation. The message that genetically determined racial traits are responsible for athletic perfor-mance as opposed to desire, coaching and culture, reinforces racism. The effects of this racist ideology are felt far beyond the world of sports. This is precisely why it must be resisted."[41]

NOTES

1. Winston Churchill, as quoted in Robert Andrews, ed., *The Concise Columbia Dictionary of Quotations* (New York: Avon Books, 1987), p. 269.

2. Ashley Montagu, *Man's Most Dangerous Myth: The Fallacy of Race* (Walnut Creek, CA: AltaMira, 1997), p. 41.

3. Jim Rorick, "World Outdoor All-Time List—Men, 2008" *Track & Field News*, January 13, 2009, http://www.trackandfieldnews.com/lists/display_list.php ?list_id=9&sex_id=M&year=2008 (accessed September 9, 2009).

4. Track & Field News World Rankings, "All-Time World Rankings—

Women's High Jump, 2002," *Track & Field News*, 2002, http://www.trackandfield news.com/rankings/women/hjworldranking.pdf (accessed September 9, 2009).

5. International Association of Athletics Federations, "IAAF World List—High Jump 2008," February 20, 2009, 2008, www.iaaf.org/statistics/toplists/inout=O/age=N/season=2008/sex=M/all=n/legal=A/disc=HJ/detail.html (accessed September 9, 2009).

6. List of high jump gold medalists available on DatabaseOlympics.com, 2009, www.databaseolympics.com/sport/sportevent.htm?sp=ATH&enum=260 (accessed September 9, 2009).

7. BBC, "Kenyan Finishes Last, but Wins Respect," BBC News, February 12, 1998, http://news.bbc.co.uk/2/hi/sport/winter_olympics_98/cross_country _skiing/55856.stm (accessed September 9, 2009).

8. See list of nations that have won Olympic medals, http://en.wikipedia .org/wiki/All-time_Olympic_Games_medal_count (accessed September 9, 2009).

9. See list of greatest Olympians based on most medals earned, http://en .wikipedia.org/wiki/List_of_multiple_Olympic_gold_medalists (accessed September 9, 2009).

10. Joseph L. Graves, *Encyclopedia or Race and Racism* (New York: Macmillan, 2007), p. 147.

11. John Hoberman, interview by the author, April 9, 2009.

12. Mike Barrowman, interview by the author.

13. Jonathan Marks, interview by the author, January 25, 2009.

14. Michael Jordan, *Driven from Within* (New York: Atria, 2006), p. 33.

15. Ibid., p. 8.

16. Graves, *Encyclopedia of Race and Racism*, p. 163.

17. Joseph L. Graves, *The Race Myth: Why We Pretend Race Exists in America* (New York: Dutton, 2004), p. 153.

18. Joseph L. Graves, interview by the author, January 23, 2009.

19. Ibid.

20. Graves, *Encyclopedia of Race and Racism*, p. 145.

21. Ibid., p. 146.

22. Ibid.

23. Ibid., pp. 146–47.

24. Ibid., p. 147.

25. Ibid.

26. Curtis W. Wienker, interview by the author, January 23, 2009.

27. S. Alexander Haslam, Jessica Salvatore, Thomas Kessler, and Stephen D. Reicher, "How Stereotyping Yourself Contributes to Your Success (or Failure)," *Scientific American*, April 2008, http://www.sciam.com/article.cfm?id=how-stereo typing-yourself-contributes-to-success (accessed September 9, 2009).

28. Gary Smith, "The Color of Speed," *Sports Illustrated*, December 6, 2004, http://vault.sportsillustrated.cnn.com/vault/article/magazine/MAG1114431/ 11/index.htm (accessed September 9, 2009).

29. Bill Brubaker, "It's Not as Simple as A . . . B . . . C," *Sports Illustrated*, August 29, 1983, http://vault.sportsillustrated.cnn.com/vault/article/magazine/MAG 1121160/ index.htm. See also Diana Nyad, "Views of Sport: How Illiteracy Makes Athletes Run," *New York Times*, May 28, 1989.

30. The Institute for Diversity and Ethics in Sport (TIDES), "Keeping Score When It Counts: Assessing the 2008–2009 Bowl-Bound College Football Teams— Academic Performance Improves but Race Still Matters," *Tides*, December 8, 2009, http://www.tidesport.org/GradRates/2008-09_Bowl_APR_GSR_Study .pdf (accessed October 5, 2009).

31. John Hoberman, *Darwin's Athletes* (Boston: Mariner, 1997), p. 4.

32. Ibid., p. 17.

33. Ibid., p. 6.

34. Ibid., p. 7.

35. Ibid., p. 8.

36. Ibid.

37. Hoberman, interview.

38. Robert A. Scott, Evelina Georgiades, Richard H. Wilson, Will H. Goodwin, Bezabhe Wolde, and Yannis P. Pitsiladis, "Demographic Characteristics of Elite Ethiopian Endurance Runners," *Medicine and Science in Sports and Exercise* 35 (2003), http://cat.inist.fr/?aModele=afficheN&cpsidt=15178034 (accessed September 9, 2009).

39. Robert A. Scott and Yannis P. Pitsiladis, "Genetics and the Success of East African Distance Runners," *International SportMed Journal* 7, no. 3 (2006), www .fims.org/default.asp?pageID=782860264 (accessed September 9, 2009).

40. Bruce Hamilton, "East African Running Dominance: What Is behind It?" *British Journal of Sports Medicine* 34 (October 2000): 393–94.

41. Graves, *The Race Myth*, p. 165.

Chapter 5

ARE YOU A PATIENT OR A COLOR?

"Race" or "ethnicity" is an inadequate proxy for the subset of human populations that are likely to benefit from a certain drug.
—Charles N. Rotimi, director, NIH Intramural Center for Genomics and Health Disparities[1]

The implications of geneticizing disease and linking health disparities to race carries with it many risks such as reification of race; belief in genetic inferiority; reaffirming inherent inequity; potentially exploiting people of color due to the market interest in this area of medicine; and the possibility of a loss of funding geared toward the prevention of disease.
—Jamie D. Brooks and Meredith L. King[2]

Race does not exist. But it does kill people.
—Colette Guillaumin[3]

"**P**hysicians increasingly use their patients' skin color or other physiological features as a first step towards 'individual' treatment, under the assumption that specific traits cluster by race," states a paper in the European Molecular Biology Organisation's journal.[4] Is this a good idea? Is there a place for race in healthcare? If so, what is the proper role for it? Are some races naturally more disease prone and unhealthy than others? Should race be a key factor in the minds of doctors when they are diagnosing and treating health problems? This is a debate with important consequences. Some proponents of the idea that race should be a key factor say that drugs can and should be designed and marketed with a specific race in mind. They point to the fact that many diseases are linked to genes that can be identified and may be present in one race or another. Combine this with the fact that members of some races suffer from some diseases at a disproportionately higher rate than members of other races and it is easy to see why race-specific drugs might be an attractive idea. But is this idea really as promising as some say? Is racial profiling appropriate in doctors' offices? And if it is, what does this say about the larger question of whether or not races are meaningful biological categories?

While health experts debate the issue, some pharmaceutical companies are not wasting any time. They are forging ahead, pushing hard to usher in the age of racialized medicine. In 2005, BiDil—a drug aimed at black people with heart problems—was approved by the US Food and Drug Administration. It was the first drug marketed for a specific race. BiDil will have lots of company in the near future, it seems. According to a 2007 report by the Pharmaceutical Research and Manufacturers Association of America, no less than 691 drugs that will target African Americans are currently in the process of being developed by US companies.[5]

This raises a slew of questions. Are people who are within racial groups really genetically alike enough to warrant all this? Do race-based drugs work as advertised? Can they help close the health gaps that currently exist between races? While such drugs may sound promising and sensible, there are some important points to consider before boarding this bandwagon.

First, however, it is important to be clear that challenging or opposing racialized medicine should not be misinterpreted as a lack a of concern for anyone's life or the very serious health disparities that exist between race groups today. I know it can't last forever, but I've been fortunate enough to enjoy extraordinarily good health throughout my life so far. There was, however, that one time when I crashed and burned. Several years ago, I came down with pneumonia and it opened my eyes about serious illness. Lying flat on my back in a hospital, waylaid by merciless germs and roasting with fever, I was forced to let go of my illusions of invincibility and to realize how vulnerable I really was. I have never forgotten how horrible and desperate I felt. I would have eagerly taken any medicine if it was a product of medical science and had a credible record of success behind it. I would not have flinched at accepting a race-based diagnosis and taking race-specific medicine if it was scientifically sound. I would not have cared if it seemed to contradict my ideas about biological race categories. When one is seriously ill, life and health are more important than winning debates or maintaining a consistent worldview.

I do not have a callous disregard for the health problems of others, especially some race groups who suffer medical problems at a higher rate than others. Citing problems with race-based medicine must not be seen as maintaining a stubborn and irrational stance against a possible cure or treatment simply because it relies on race. Sensible, good-hearted people who oppose race-targeted medicine do so because they are not convinced that the science and logic behind it are sound. No decent right-thinking person would want to slow down or prevent medical treatment that would ease suffering or save lives, even if it did contradict a deeply held position. As this chapter will show, there are many problems with racialized medicine. Nothing you will read here, however, should suggest to you that I or anyone I quote would ever put ideas about race before a human life. If racialized medicine works, then bring it on, I say. But we have to be sure that creating and marketing drugs for specific races is really the best way forward. Far too many lives are at stake to charge at full speed down the wrong path.

While it is not a direct argument against race-based medicine, it should be acknowledged up front that medical science has a tainted past regarding race. There have been terrible episodes, including racist craniometry and phrenology, the attempts to rank races based on skull shapes and sizes; eugenics, the idea that humankind could be improved by eliminating or reducing through sterilization undesirable people and races; and the Tuskegee syphilis study (1932–72), in which treatment was withheld from black men with syphilis in order to give the researchers an opportunity to learn about the long-term effects of the disease. In many tragic instances, race-based medical practices have led to suffering and death. In contrast, some examples of the ways racism can pervert medical science appear laughable today. My favorite is "drapetomania." This was a creative —and convenient—"disorder" that some nineteenth-century white American doctors assigned to black slaves who escaped or attempted to escape. A black person who suffered from this "disorder" had failed to accept his or her "natural state of servitude." It was, however, a treatable condition, according to the experts of the day.[6]

While past mistakes and crimes perpetrated by racist scientists and doctors must be acknowledged, remembered, and guarded against in the future, history is not the decisive factor when it comes to the question of whether or not race-based medicine is useful today. It could well be that, yes, race and medicine were a bad mix in previous centuries, but now race and medicine are coming together in a new way that will benefit people. However, as we shall see, there are many problems to look out for—so many, in fact, that perhaps racialized medicine is not the best way to address health problems.

I KNOW WHAT DRUGS ARE, BUT WHAT ARE RACES?

As you might expect, the first problem with race-based medicine is linked to the core problem with the concept of race itself. Biological races simply do not exist as the convenient subspecies of humanity that most people imagine them to be. There are no firm and consistent categories of races based on a common biology. Yes, there may be a similar genetic ancestry shared by members of a traditional race group. (African Americans may share some recent African ancestry, more than other Americans who have more distant African ancestry.) But the variation can be great. Actress Halle Berry has a white mother and a black father but is identified as black or African American in the United States, just as the American-born daughter of two recent immigrants from Somalia would be. Race usually comes down to where you are, the cultural rules that apply at the time, and the "race" of at least one of your biological parents. A person's true genetic makeup and deep ancestral history is usually subordinate to those factors. Clearly this is a huge problem for race-based medicine, because healthcare must be about a person's real biology rather than the fleeting, flimsy, and inconsistent rules of culturally determined races. How, for example, can we have a "white person's drug" when the definition of a white person can change at the whim of culture? (Don't forget that Italian, Greek, Polish, and Irish immigrants did not qualify as "white people" just a short time ago in America.) If I am sick, I don't want my doctor making assumptions and drawing conclusions based on my race. Why? Because, when it comes to my genetic history, I don't really know for sure what or who I am, so how can my doctor possibly know? The only thing I'm sure of is that I'm a descendant of prehistoric Africans who lived about 100,000 years ago; between then and now, however, things become a little fuzzy to say the least. I may look "white" and I may have been assigned to the white race at birth, but that doesn't necessarily tell the whole story or the true story. I could have had a great-great-grandparent who was "black" and another who was Native American. Given the realities of American history, especially in the south, where both of my parents have family roots, it is possible. Given the chance, I would prefer to have my personal genome analyzed and let my doctor take it from there. Barring that, I feel it would be best for the doctor to think of me simply as an individual human being because that's about the only thing we can be absolutely sure of. I don't want to wait for the era of personal genome sequencing. I want to be considered an individual right now! I am not the embodiment of something called the white race. I am one single expression of humankind, put together over the last two hundred thousand years and beyond. I hope I am not just being vain, but I really do think I'm a bit more complex and unique than the simple label "white man" would indicate.

My father was adopted as a child and knows little or nothing about his distant past. He has black hair so it is possible that he could have some Mediterranean ancestry, perhaps Greek or Italian. With this in mind, what do you suppose might happen to me if one day I showed up in an emergency room in America with severe stomach pains and anemia? I'm white and my parents are white. Sickle-cell disease might not cross the minds of the ER staff. It might take a dangerously long time for the doctors to figure out the cause of my symptoms. I could suffer for a long time. Doctors might even cut me open by mistake. All because it never occurred to them that sickle-cell disease could be my problem. But why should it occur to them? That's a "black" person's disease, right? No, it's not.

"Sickle cell is a disease of populations originating from areas with a high incidence of malaria," writes race researcher Kenan Malik in an essay about race and medicine. "Some of these populations are black, some are not. The sickle-cell gene is found in equatorial Africa, parts of southern Europe, southern Turkey, parts of the Middle East and much of central India. Most people, however, only know that African Americans suffer disproportionately from the trait. And, given popular ideas about race, they automatically assume that what applies to black Americans also applies to all blacks and only to blacks. It is the social imagination, not the biological reality, of race that turns sickle cell into a black disease."[7]

Sickle-cell anemia is found in many people who are not identified as black. For example, it appears in some nonblack Hispanic people, some people in northwest India, and many people throughout the Mediterranean region. "The label 'black disease,' however, rendered the distribution of sickle-cell anemia invisible in other populations leading to erroneous understanding of the geographical distribution of the underlying genetic variants. This is one reason why many people, including physicians, are unaware that the town of Orchomenos in central Greece has a rate of sickle-cell anemia that is twice that of African Americans and that black South Africans do not carry the sickle-cell trait."[8]

Oh, and my imaginary scenario of being a white person with a "black disease"? It's not so imaginary. I found an article in the medical journal *Annals of Internal Medicine* that relates the case of an eight-year-old boy who suffered the symptoms of acute sickle-cell disease. Because he looked "European" to doctors, however, he was almost put through unnecessary exploratory surgery. The doctors caught their mistake just in the nick of time.[9]

Race belief constantly tells us otherwise, but we have to remember that physical characteristics can make people seem more or less alike than they really are. A "mixed-race" couple in Great Britain made news in 2009 when they produced their second set of twins in which one sibling appeared white and the other black.[10] What if a racially profiling doctor gave the fraternal sisters different race-based medications because they belong to different "races"? I recall that during a visit to Fiji, I was often amazed at how much some indigenous Fijians looked like some Caymanians. (I have lived in the Cayman Islands for many years and know the people well. The Caymanian population is highly diverse, but there are some specific physical traits that pop up frequently.) The physical similarity between these people on opposite sides of the world was remarkable. They might speak differently, dress differently, eat different foods, and listen to different music, but in many cases their facial features, skin color, and hair type were a near perfect match. It would certainly be easy for someone to think of them as the same race. What if this was to be done in a medical context? Who knows? It could be disastrous. Fijians and Caymanians are not only far apart geographically, but they are also far apart genetically, relative to the entire species. If a doctor is going to base decisions on genes, then those decisions must be based on genes, not on superficial and inconsistent clues such as skin color, facial features, and hair type.

Race-based medicine has a very high mountain to climb when it comes to the genetic diversity of African Americans. This category includes millions of people lumped together on the basis of little more than whether or not they had at least one parent who was black, rather than on the basis of their actual genetic history. African Americans include a wide variety of people. The one-drop rule, which declares that any black ancestry makes you black, has worked well over the last few centuries to ensure that this group of Americans is far from homogeneous. For example, President Obama identifies himself as a black man and an African American, which culturally he is. But, if he has health issues in the future, should his doctor rely on that label and prescribe some new race-based drug that was made for black people?

Yes, President Obama is an African American, but prescribing "African American drugs" for him might be dangerous. Although he qualifies as an African American—he is an American citizen and his late father was a dark-skinned Kenyan—President Obama is not the usual sort of African American, and this is not just because he lives in the White House. Most African Americans have ancestors who were taken to America as slaves from western and central Africa. Kenya, where Obama's father lived, is in East Africa. Given the fact that Africa has the greatest genetic diversity in the world, Obama's genetic history is likely to be significantly different from that of a typical African American. But in the eyes of Americans, including, perhaps, the eyes of doctors, Obama is simply an African American or black man. The fact that he had a white European American mother and a black East African father—his true genetic history—might easily be overlooked or ignored, despite the fact that it could be relevant to diagnosis or treatment. So it is not difficult to imagine Obama or any other

African Americans with an atypical genetic history being prescribed medication that was designed not for them but for more typical African Americans, those who trace most of their ancestry to western or central Africa.

It is important to remember that there is more genetic diversity within races than there is between races. Additionally, the race categories are not firm or rigid in any way. Every time someone is born, dies, or makes a baby, the gene pool changes and so do race groups, whether people are aware of it or not. Assigning someone to a race based on skin color, facial features, and hair type may feel like an objective decision based on that person's genetic history, but it really is not. It is a decision influenced—if not determined by—one's particular cultural perspective. African American or black, Asian, Native American, Hispanic, and white are labels given to massive groups of diverse people around whom American culture has drawn borders. Never forget that in different time periods and in different cultures the rules are rarely if ever the same. Basing the design and targeting of new drugs on race categories is risky to say the least. Something is likely to go wrong simply because racial borders are so subjective, inconsistent, and unreliable.

Biological anthropologist Jonathan Marks has argued for many years against claims that races are natural and useful categories. Dr. Marks specializes in molecular genetics and strongly disagrees with any geneticist who promotes the idea that biological races are sensible categories. Dr. Marks cites the basic problem of using race to make assumptions about an individual's real genes and genetic history. But he is also concerned about the dangers of race-targeted drugs working as advertised for some but not all members of a race, as well as potentially working for a significant number of people outside the specified group who will never get these drugs. The whole business is simply too sloppy, he feels, and could have severe consequences.[11]

"We know race is a terrible proxy for genotype," explains Dr. Marks. "The most relevant examples would be something like a drug working in 75 percent of Africans but 25 percent of Asians. First, within each category, there will have to be huge variation, patterned the way we know human diversity is patterned: The African Ethiopians are more likely to cluster with the Asian Pakistanis than with the African Ghanaians. So the continental average value is hardly of use. More importantly, the drug won't work in 25 percent of Africans—and might sicken them—and will work in 25 percent of Asians [who won't receive its benefits because it won't be given to them]. What I'm getting at is that the test has to be made at the individual genotype level; prescribing a drug based on the census category of the patient is a medically very bad idea."[12]

Race-based medicine is an "inappropriate and perilous approach" to healthcare, argues bioethics expert Sharona Hoffman in the *American University Law Review*:

> The argument is rooted partly in the fact that the concept of "race" is elusive and has no reliable definition in medical science, the social sciences, and the law. Does "race" mean color, national origin, continent of origin, culture, or something else? What about the millions of Americans who are of mixed ancestral origins—to what "race" do they belong? To the extent that "race" means "color" in colloquial parlance, should physicians decide what testing to conduct or treatment to provide based simply on their visual judgment of the patient's skin tone? "Race," consequently, does not constitute a valid and sensible foundation for research or therapeutic decision making.
>
> Further … "racial profiling" in medicine can be dangerous to public health and welfare. A focus on "race," whatever its meaning in the physician's eye, can lead to medical mistakes if the doctor misjudges the patient's ancestral identity or fails to recall that a particular condition affects several vulnerable groups and not just one "race." The phenomenon can also lead to stigmatization and discrimination in the workplace and elsewhere if the public perceives certain "races" as more diseased or more difficult to treat than others. In addition, "racial profiling" could exacerbate health disparities by creating opportunities for health professionals to specialize in treating only one "race" or to provide different and inferior treatment to certain minorities. It could also intensify African Americans' distrust of the medical profession. Finally, "race-based" medicine might violate numerous anti-discrimination provisions contained in federal law, state law, and federal research regulations and guidelines.[13]

Another problem for race-based medicine is that many people are just not comfortable talking about race and openly basing decisions on race. This is understandable, given the history and emotions that come with that heavy word, "race." Clyde Yancy, president of the American Heart Association and medical director of the Baylor Heart and Vascular Institute, says

BiDil, the drug targeted at African Americans, has had very low prescription rates. "The general awkwardness surrounding racial issues in our society bleeds into medicine," explains Yancy. "There may be unique mechanisms at play in heart failure in some people described as African American. When a practitioner is presented with an African American patient, [he or she] may be hesitant to offer the patient a drug based on race. And some patients are put off when practitioners emphasize race.... We need to move away from race quickly. As we mature, we will be able to supplant the notion of race as predictor of response with something more palatable to the scientific community and to patients. Then we don't have to bring the word heft of 'race' into how best to care for patients."[14]

In 2009, journalist Jerry Adler wrote an article in *Newsweek* about BiDil and the broader issue of race-based drugs. He noted that "the very idea of marketing a drug that way provoked outrage among people who saw it as a validation of the concept of 'race' as a genetic, rather than a social, category. Perhaps because of that controversy, as well as other reasons, including the drug's high cost, it was recently estimated that only 3 percent of the patients who might benefit from BiDil were actually getting it."[15]

Adler quotes Timothy Caufield, a professor of health law at the University of Alberta, who believes that race-based medicine is problematic because race simply is not a very useful biological category: "Someone whose ancestors came from Nigeria is very different from a descendant of Kenyans, but if the two of them are walking down a sidewalk in New York, they're both 'black.' You can try to make those distinctions in your research, but once it gets into the hands of the drug manufacturers, there's going to be slippage.... marketers want to sell to the broadest possible categories."[16]

DIAGNOSE ME, NOT MY RACE

When James Watson, a codiscoverer of the structure of DNA, had his genome sequenced in 2007, it cost about a million dollars. In early 2009, *New Scientist* reported that a company had sequenced the genome of a Nigerian man for under $60,000. The same report included a claim by a company named Complete Genomics that it could do it for a mere $5,000.[17]

Dr. Rasmus Nielsen, a University of California, Berkeley, biologist, predicts the cost of genome sequencing will soon drop to a few thousand dollars and eventually to a few hundred dollars. It's coming and it's coming fast, he predicts. "You will have your genome sequenced, and I will have my genome sequenced."[18]

So, while we aren't quite there yet, it seems inevitable that truly personalized healthcare, which includes the real genetic makeup and histories of patients, will soon be available on a mass scale. (Well, at least in the wealthy nations.) This will revolutionize healthcare, and it seems logical that it will eliminate any need for race-based medicine. Why be a race when you can be you? All the traditional notions about race will be irrelevant. It won't matter who you or your doctor think you are or who you assume your genetic kin have been over the last several thousand years. The only things that will count will be what your genome says, your lifestyle, your diet, and your symptoms if you are ill. This will be the overwhelming reality that will trump race belief when it comes to making decisions about keeping you alive and healthy.

For years, futurist and inventor Ray Kurzweil has been saying that Moore's Law, the exponential shrinking of transistors enabling more computer power to be packed into smaller and smaller spaces, combined with the fast-rising fields of nanotechnology, robotics, and genomics, promises a flood of stunning new abilities for humankind. "When the human-genome scan got underway in 1990 critics pointed out that given the speed with which the genome could then be scanned, it would take thousands of years to finish the project," Kurzweil points out. "Yet the fifteen-year project was completed slightly ahead of schedule, with a first draft in 2003. The cost of DNA sequencing came down from about ten dollars per base pair in 1990 to a couple of pennies in 2004 and is rapidly continuing to fall."[19]

Newsweek's Adler closed his essay on race-based medicine by—you guessed it—bringing up the promise of individualized medicine: "The research that has allowed us to parcel out racial differences by ancestry will eventually outstrip even those categories, and identify specific vulnerabilities and drug reactions in the genomes of individuals. We will no longer be white or Asian or African, or even Northern European, Ashkenazi, Japanese, or East African; we will be who we are, each one of us. And the sooner we reach that point, the better."[20]

Amen to that. I would add, however, that we should all be treated as individuals *right now*, no matter whether science is ready for it or not, no matter whether we can afford to have our personal genome sequenced or not. Again, a person may appear to be a run-of-the-mill white American or black American, for example, but who knows what surprising truth may lie beneath the surface? I would not want to be given a drug that was tailor-made for white Europeans exclusively, just because I look like a white European. I would be worried that something relevant might be lurking in my DNA that my doctor and I didn't know about. Nor would I want to miss out on a drug that might help me, just because my doctor neglected to tell me about it, on racial grounds.

BIOLOGICAL RACES MAY NOT BE REAL, BUT RACIAL PROBLEMS ARE

Perhaps the greatest problem with race-based medicine is that it can distract us from the need to find solutions to serious race-based health problems. If medical researchers promise that they can reduce the numbers of black people who suffer from heart attacks and diabetes with a "black drug," for example, then maybe some of these researchers, as well as politicians and doctors, will feel they don't need to worry about the underlying causes of high rates of heart disease and diabetes. Don't worry about the ongoing problem of residential segregation and lack of access to good healthcare for some racial groups. There will be no need to be concerned about the large numbers of black people who are uninsured and, as a result, rarely visit doctors and get checkups. Why be troubled by the problems of racism, poverty, nutrition, education, employment, and so on? Just pop a race-based pill and everything will be better.

Law and bioethics expert Pilar Ossorio points out that one can "make money on medications that lower blood pressure, especially if people are going to be taking that medication for the rest of their lives. But how does a company monetize the eradication of racism and poverty?" [21]

Race affects people in powerful, direct ways that have little or nothing to do with genetic history. For example, David R. Williams of Harvard, a leading expert on health disparities, points to ways in which socioeconomic status intersects with race to impact the health of Americans:

> We know that, across the world, socioeconomic status is a powerful predictor of health. In fact, it's a more powerful predictor of health than

genetics or medical care or cigarette smoking. Now, why is that important to our discussion of race? It's very important because, on average, in
our society [the United States], socioeconomic status differs by race. So,
on average, blacks have lower levels of income, lower levels of wealth,
and lower levels of education than whites do. And for other minority
populations, a similar pattern is evident. So, many of the racial differences we see are not due to skin color, but to the fact that disadvantaged
minority populations have fewer economic resources than whites do. But
that certainly does not account for all of them. What we find across multiple measures, or multiple indicators of health status, is that at every
level of economic status, blacks are still doing more poorly than whites.[22]

The people popularly known as black, Native American, and Hispanic
are thought by whites to be neatly divided up according to naturally occurring biological categories. They are not. However, there are many biological consequences that come with membership in those groups. Confused?
Try this thought experiment. Imagine a million people who are members
of the "Purple Club" and a million people who are members of the "Green
Club." All two million are genetically similar and are similar in their current general health. Now imagine treating the Green Club members
poorly by limiting them to low-paying jobs, inferior schools, and inferior
healthcare. Also add to their stress by constantly communicating to them
that they are not as smart or morally upright as the members of the Purple
Club. Do this for a couple of hundred years or so and then compare the
health of the members of the two clubs. Is there any doubt that the Green
Club members would probably have more serious health issues than the
Purple Club members? How could they not? The question is, why would
anyone want to focus on genetic differences based on club membership to
find a reason for the health disparity? Wouldn't very different life experiences be the obvious first place to look?

When it comes to racial health disparities, the ultimate goal should not
be to cure enough black people that the numbers begin to line up favor-
ably with those for white people. The intelligent and moral goal is to fix
the social problems that probably lead to the racial health disparities in the
first place. Doctors love to say that prevention is better than cure. We
cannot let the promise of race-based drugs allow us to assume that they are
the solution to important social challenges. Drugs will never cure the ills
brought upon our societies by racism.

"There is more genetic variation within each race than between races," says Harvard public health expert Williams:

> In other words, some black people are more similar genetically to some white people than they are to other black people. So, genetic differences are not plausible from a scientific point of view to account for this really striking pattern of racial disparities that we see across the fifteen leading causes of death. We need to understand what it is within the social environment that produces these patterns of ill health that exist for multiple health conditions and that have been so stable and consistent over time.
>
> I am not saying we shouldn't study genetics at all. Genetics plays a role, but it plays a role in interaction with the environment and is most likely a minor contributor to the pattern of health disparities that we see.[23]

The 2009 report, "Geneticizing Disease: Implications for Racial Health Disparities," written by Jamie D. Brooks and Meredith L. King for the Center for American Progress, warns of getting it wrong when apportioning blame for the comparatively poor health of blacks, Native Americans, and Hispanic Americans: "Racial and ethnic minorities bear the burden of poor health and health outcomes at a much higher rate than whites in the United States. The danger, though, lies in extrapolating from these facts that race as a genetic factor has something to do with health disparities."[24]

The impact of race (as in the culturally determined membership of a social group) upon health can be deceptive, because it may play out over a lifetime in ways that can obscure its influence. Public health expert Williams warns against thinking of economic status only in terms of a person's current status. The person's entire lifetime must be considered, he says, because a middle-class or wealthy person may have been poor in childhood and exposed to a variety of negative environmental factors as a result. This could mean such a person is at a higher risk for some health problems than her or his present socioeconomic status suggests.[25]

Racism still exists in the United States and around the world. While significant progress has been made, it clearly remains a problem for many people in many places. Hopefully this much is obvious to everyone. What may not be obvious is the stress that racism causes for those on the receiving end of it and how it can degrade their health. First, it is important to understand that racism does not come only in the form of hate groups firebombing houses or yelling racial slurs. Racism also creeps into the subconscious thoughts of those who find themselves cast as inferior by a dominant category of people. Once in the subconscious, it can wreak quiet havoc on the body and mind. This "internalized racism" plays a significant role in health disparities, according to Dr. Williams. "Where some groups are regarded as dominant and some are subordinate and some groups, based on race, are viewed negatively, some proportion of persons who are viewed negatively will actually buy into and believe society's negative characterization of them. When individuals do that, it is an example of internalized racism—they have accepted as true the society's negative characterization of their group. And there's research that suggests that all of those dimensions impact on health."[26]

It has been found that "learned helplessness could be a powerful force shaping the psychological orientation of individuals, and it could actually shape racial depression," according to Dr. Williams:

> When individuals in their early lives have always experienced failure, have always experienced blocked opportunity, sometimes even when the door of opportunity opens they have learned to be helpless. So, one of our challenges is to make real to individuals that there are opportunities, and also to create opportunities that provide a ladder out of their difficult situation so that they do not fall into this trap of learned helplessness.... Following individuals over time, research has found that those who are hopeless, who don't have a future to believe in, have more rapid development of heart disease as measured in their blood vessels. It's common sense that at some level it would affect us psychologically, but what that research suggests is that hopelessness—not being able to believe in a future—is actually killing us. It's having negative physiological consequences for how we are able to function.[27]

Environment matters. If you live in a rundown neighborhood with plenty of liquor stores and junk food available but little or nothing in the way of fresh fruits and vegetables to be found, your health may suffer as a result. If you constantly worry about violent criminals on the streets around your home, you may not be very interested in jogging or walking in the evenings, and you may end up overweight and unfit as a result. The frustration of not getting a needed raise, possibly because of a racist boss, combined with a less than adequate income might wear on some people over the years and lead to a health crisis one day. If that health crisis comes, a race-based pill will be welcomed if it helps. But that pill won't put healthy food into shops in poor neighborhoods, stop crime, or change the heart of a racist boss.

It seems clear that the massive health deficiencies of nonwhite groups in America are primarily if not entirely environmental. There is no reason to think that black people, Native Americans, and other groups in America are prone to more illnesses and die younger than white Americans because they are genetically inferior. A far more reasonable and likely explanation is that the United States remains under the long shadow of slavery, residential segregation, and racism. It may seem like ancient history to some, but the echoes of America's past racial injustices are still there. This is not to suggest that all challenges facing nonwhite people in America come directly from the dominant group. There are serious problems within contemporary black American culture that probably play a role as well.[28] Who can doubt, for example, that it would improve conditions for many people if larger numbers of low-income black American families adopted a culture revering education in the way that many Asian American families do?

The cold hard numbers paint a stark picture of how different races experience life differently, even when sharing space in the same society. Today in the United States, white Americans can expect to live, on average, 78.3 years; for black Americans, however, it's just 73.1 years. Mexican Americans and black Americans are 67 percent and 60 percent, respectively, more likely to suffer from diabetes than white people are. They are also more likely to die from it than white people. And probably the most disturbing statistic of all is this: black babies are two and a half times more likely to die than white babies in America.[29]

Again, to emphasize the point, there is nothing to suggest that the black babies who die are genetically (racially) inferior to white babies. Race—as in a genetic or biological category—does not doom them. Race —as in culture—does. They are victims of a country and a world divided up into races, a system that inevitably leaves too many people holding the short straw.

University of North Carolina epidemiologist Jay Kaufman states:

The important question for health disparities is whether there are differences in the mean for different populations. And there's no reason to think that any one continental population would have more of whatever genetic trait it is that might allow you to live a longer life. There's no reason to think that the average [life expectancy] for the population of Africa would be higher or lower than that of Europe. We do see variability within a population, tremendous variability. But disparities are about differences between populations. And there's no consistent evolutionary theory, nor is there any evidence, that suggests that genetic determinants explain any of the mortality differences found between populations.[30]

One very telling long-term study sheds light on the differences in birth weight of babies born to black and white mothers in America. This one really got my attention. Low birth weight is a serious health risk, and black babies face this problem at a higher rate than white babies in America. Surprisingly—or not, if you don't put much stock in the concept of biological races—researchers found that babies born to African women who had immigrated to the United States were the same weight as white American babies. But the daughters of these immigrant African mothers later gave birth to babies who were, on average, half a pound lighter than the babies born to white women or their own mothers.[31]

The researchers also found that black women who immigrated to the United States from the Caribbean gave birth to infants who weighed more than the babies of American-born black women. "Something is driving this that's related to the social milieu that African American women live in throughout their entire life," said one of the scientists who conducted the study.[32] Yes, it appears that there is something about simply living your entire life as a black woman in America that makes you more likely to have a baby that weighs less and is at greater risk. If it were a genetic "race" problem, then the black African and Caribbean immigrants should also give birth to lower-weight babies in America. But they don't. The reason is not in the genes of black American women. The problem is in the society they live in.

Another glaring example of racial health disparities that seem to be about race-culture rather than race-genetics is the unfortunate health troubles of the Pima Indians in Arizona. Nearly half of the adults have type 2 diabetes, possibly the highest rate in the world. But the Pima Indians just across the border in Mexico have a type 2 diabetes rate of about 7 percent.[33] It makes no sense that two groups of people, essentially the same genetically, would exhibit such glaring health differences for anything other than environmental reasons. It is unlikely that a race-based pill is what the American Pima Indians need. They need a solution to the social conditions that are negatively impacting their health.

The reality of an unfair world—including a race-based lifelong health burden or even a death sentence—can shock and anger compassionate people. One senses that something is terribly and inexcusably wrong when a child suffers and dies, apparently because of nothing more than the child's racial identity. Clearly the United States has a moral obligation to do more. Black babies dying at a rate more than twice that of white babies is a crisis, and a solution should be sought immediately. But this is not just an American problem.

An estimated five hundred thousand women worldwide die from complications related to pregnancy and childbirth each year. Incredibly, half of these deaths occur in sub-Saharan Africa, a region with less than one-sixth of the total global population. Each day in the developing world, some twenty-five thousand children under the age of five die from malnutrition and diseases that are easily preventable or treatable in wealthy nations. That adds up to about nine million children per year, some *ninety million per decade*. Virtually all of these lost babies are nonwhite. What kills them, other than the fact that they have the bad luck to be born into extreme poverty? They die because on the day of their birth they inherit deadly, dysfunctional environments marked by malnutrition, contaminated water, lack of good sanitation facilities, lack of vaccines, and lack of access to doctors. Does anyone honestly believe that we need a special race-based drug to save these ninety million black- and brown-skinned babies who keep dying decade after decade? Should the focus be on their genes and "race" or should it be on their diets, their living conditions, and the quality of healthcare they receive? The conditions they live in kill them, not their racial genetic makeup. It is not a pill they need; they need to be born into better environments. And until we, the adults, decide to make it better, the babies will keep dying.

Hypertension (high blood pressure) is a significant health problem in the United States, afflicting one in five people. Once again, nonwhites tend to be at greater risk. American Indians and Native Hawaiians are diagnosed at a rate of one in four. But for African Americans the rate is one in three.[34] A revealing study by epidemiologist Michael Klag found that the rate of hypertension among African Americans rises with the darkness of their skin. The darker the skin, the more likely hypertension is to be a problem. However, this only holds true for black people of lower socioeconomic status. Skin color does not predict hypertension rates among black people of higher socioeconomic status. This seems to go against what would be expected if a genetic-racial component were solely to blame for the high rate of black hypertension in America. "The results of the study suggest that a genetic mechanism for high blood pressure is less likely or, if there is a genetic tendency, it is not the biggest factor," Dr. Klag told the *New York Times*. "The study means we have to concentrate on environmental factors that may be causing hypertension, not color."[35]

Those who tend to think of hypertension in terms of race need to be aware of another important point. Black rates of hypertension vary not just within the United States by race but also beyond America's borders. Hypertension rates for British and Caribbean people of African descent are two to three times lower than they are for people of African descent in the United States. And, guess who have the highest rates of hypertension in the world? White people in Germany. It is clear, states the medical report, "Geneticizing Disease: Implications for Racial Health Disparities," that "the hypertension health disparity suffered *by* blacks and even *within* the black population compared with white Americans has social origins and biological effects."[36]

It would be so easy to declare that race as a concept is too inconsistent and confusing to be used in the creation and distribution of drugs or in anything else to do with healthcare. But we must acknowledge that we live in a racialized world. As long as people think of themselves and others as members of racial groups, and treat one another differently as a result, there is a need to consider race in healthcare. It doesn't matter that races do not hold up as biological categories of people. The fact remains that race is a powerful cultural reality affecting the health of millions of people. To say that everyone in the United States and throughout the world must receive the same high-quality healthcare that is at present available to the wealthiest people may be unrealistic and naïve given current political and economic realities. But that does not relieve us of our moral obligation to reach for that goal. If we want nonwhite babies to have the same chance to live and grow as white babies do, then we need to do the hard work that can make it happen.

Wealth and health are inextricably connected in the United States. Many experts believe that the income gap between blacks and whites is the primary reason for the gap in health outcomes. In 2007, the median income for white households in America was $54,920, while the median income for black households was $33,916. Non-Hispanic whites had a poverty rate of 8.2 percent in 2007. Hispanics had a rate of 21.5 percent and blacks had a rate of 24.5 percent.[37] The percentage of uninsured black Americans was nearly double that of non-Hispanic white Americans in 2007, at 10.4 percent versus 19.5 percent.[38] Could this be why white Americans live longer on average than black Americans?

While conducting research for this book, I was surprised and disturbed to discover just how severe the racial health disparities are in the United States. I knew about African Americans' problems with hypertension and was aware that black men had higher rates of prostate cancer, but I had no idea how long the list of black/white health inequalities was. Consider these statistics from the American Lung Association:

- Between 2000 and 2003, African American men were 37 percent more likely to develop lung cancer than white men;
- Lung cancer incidence rates among African American women are equal to that of white women, although rates of smoking in African American women are much lower;
- Between 2000 and 2003, the lung cancer mortality rate in the African American population was 12 percent higher than that of the white population;
- Between 1996 and 2002, the lung cancer survival rate was only 12.8 percent for African Americans while it was 15.8 percent for whites.[39]

The American Cancer Society reported in 2007 that while the racial disparity between blacks and whites had decreased, the death rate for all cancers combined continued to be 35 percent higher in African American men and 18 percent higher in African American women than in white men and women: "African Americans have the highest death rate and shortest survival of any racial and ethnic group in the United States for most cancers. The causes of these inequalities are complex and interrelated, but likely arise from socioeconomic disparities in work, wealth, income, education, housing and overall standard of living, economic and social barriers to high quality cancer prevention, early detection and treatment services and the impact of racial discrimination on all of these factors. Biological or inherited differences associated with 'race' are thought to make a minor contribution to the disparate cancer burden among African Americans in the United States."[40]

And finally, these chilling words and statistics from the US Department of Health and Human Services:

- African Americans have the highest mortality rate of any racial and ethnic group for all cancers combined and for most major cancers. Death rates for all major causes of death are higher for African Americans than for whites, contributing to a lower life expectancy for both African American men and African American women.
- In 2004, African American men were 1.4 times as likely to have new cases of lung and prostate cancer, as compared to non-Hispanic white men.
- African American men were almost twice as likely to have new cases of stomach cancer as non-Hispanic white men.
- African American men had lower five-year cancer survival rates for lung and pancreatic cancer, as compared to non-Hispanic white men.
- In 2005, African American men were 2.4 times as likely to die from prostate cancer, as compared to non-Hispanic white men.
- In 2005, African American women were 10 percent less likely to have been diagnosed with breast cancer, however, they were 34 percent more likely to die from breast cancer, as compared to non-Hispanic white women.
- African American women are twice as likely to have been diagnosed with stomach cancer, and they are 2.4 times as likely to die from stomach cancer, as compared to non-Hispanic white women.[41]

So what do all these numbers really mean? Former US Surgeon General Dr. David Satcher and other health experts have tried to make them easier to comprehend and place in context. They estimated that 886,202 deaths between 1991 and 2000 were the result of health disparities between blacks and whites in America. During that same period, advances in medical technology saved 176,133 lives. That's a big difference, especially considering the amount of money that goes into new technology compared with the money spent trying to eliminate racial health disparities:

> Much of the billions of dollars spent in the United States to improve health outcomes is directed at the "technology" of care—the race among private industries and academia to develop better drugs, devices, and procedures. Far less money and infrastructure is devoted to improving health by enhancing equity—achieving equal care for equal need—and eliminating disparities in the treatment and outcomes of those with similar conditions.
>
> Resolving the causes of higher mortality rates among African Americans can save more lives than perfecting the technology of care. Policymakers could act on this information without waiting for more precise projections. The prudence of investing billions in the development of new drugs and technologies while investing only a fraction of that amount in the correction of disparities deserves reconsideration. It is an imbalance that may claim more lives than it saves.[42]

According to a 2005 report by the American Sociological Association, the problems are social (not genetic), as are the solutions:

> Research…shows that racial and ethnic differences in health outcomes stem from socioeconomic inequalities, adverse conditions in segregated neighborhoods, as well as institutional practices that favor whites over minorities. Reducing poverty, integrating neighborhoods, raising education levels, and reducing prejudice against racial/ethnic minorities would improve the likelihood of healthier and longer lives for minority groups. There is strong evidence that health insurance increases access to quality medical care and that people with medical insurance are likely to be healthier, but access to healthcare is not the whole answer. Policy-makers, civic leaders, and healthcare providers must address the lack of healthcare as well as the factors associated with extreme residential segregation (especially among African Americans) that contribute to poor health. Access to affordable healthcare, neighborhood cleanliness and safety, proximity of amenities that promote healthy lifestyles, and desegregation are among the issues that bear substantially on life or death, illness or health.[43]

Another danger that likely would come from a widespread embrace of racialized medicine is further "confirmation" of the reality of races in the minds of millions of people. "An over-reliance on gene causation as a frame for the study and treatment of disease has serious racial implications," warn the authors of "Geneticizing Disease":

> To operate under the notion that races are genetically different, with no proof to date to verify this, ignores the complexity of population genomics. This results in people using genetics as a proxy for race. The potential downfall of this is the reification of race—taking race out of the social construct in which it was created and placing it within the context of human genetics.... doing this...gives race a biologic truth or legitimacy which it does not have, thereby distorting the reality of what race actually is.
>
> And while race is being reified, a survey of the biomedical research being conducted shows race is often used as a valid classification. Within the United States, race, ethnicity, ancestry, and culture are used interchangeably. How accurate can "race" be in determining genetic links to disease and health conditions when the definition of race is one that eludes most researchers?[44]

White people in poor towns up in the Appalachian Mountains don't live as long, on average, as wealthy white people in New England states do. One West Virginia town, for example, has a life expectancy of just fifty-five years.[45] Does this mean that gene-based drugs, designed specifically for members of the "Appalachian Mountains race," are needed? Or do you think maybe the people there need social and environmental solutions to the social and environmental problems that set them up for their health problems in the first place?

It took the Human Genome Project ten years and $3 billion to sequence one genome. In the coming years, it is likely that the time and cost required will drop to a tiny fraction of that, on a par with many of today's common medical tests. When this happens, few people, if anyone, will fail to agree that a category as sloppy and inconsistent as race is not needed for the diagnosis and treatment of patients. Apart from the impact of socially constructed race on the health of people across societies and the world, I fail to see the necessity for the use of this category now. I'm no doctor, but when it comes to healthcare for individuals, I recommend skipping "race" and proceeding directly to "individual."

NOTES

1. Charles N. Rotimi, "Are Medical and Nonmedical Uses of Large-Scale Genomic Markers Conflating Genetics and 'Race'?" *Nature Genetics* 36 (2004): S43–S47, www.nature.com/ng/journal/v36/n11s/full/ng1439.html (accessed September 9, 2009).

2. Jamie D. Brooks and Meredith L. King, "Geneticizing Disease: Implications for Racial Health Disparities," Center for American Progress, 2009, p. 23, www.americanprogress.org/issues/2008/01/pdf/geneticizing_disease.pdf (accessed September 9, 2009).

3. Colette Guillaumin, *Racism, Sexism, Power, and Ideology* (London: Routledge, 1995), p. 107.

4. Katrin Weigmann, "Racial Medicine: Here to Stay?" *EMBO Reports* 7, no. 3 (2006): 246, www.nature.com/embor/journal/v7/n3/full/7400654.html (accessed September 9, 2009).

5. Pharmaceutical Research and Manufacturers of America, "Medicines in Development for Major Diseases Affecting African Americans," September 2007, www.phrma.org/files/African%20Americans%202007.pdf (accessed September 9, 2009).

6. Brooks and King, "Geneticizing Disease," p. 13.

7. Kenan Malik, "Is This the Future We Really Want? Different Drugs for Different Races," *TimesOnline*, June 18, 2005, www.timesonline.co.uk/tol/comment/columnists/guest_contributors/article534565.ece (accessed September 9, 2009).

8. Rotimi, "Are Medical and Nonmedical Uses?"

9. Ritchie Witzig, "The Medicalization of Race: Scientific Legitimization of a Flawed Social Construct," *Annals of Internal Medicine* 125, no. 8 (October 15, 1996): 675–79, www.annals.org/cgi/content/full/125/8/675 (accessed September 9, 2009).

10. Associated Press, "British Couple Has 'Black-and-White Twins' Twice," MSNBC, Health, January 2, 2009, www.msnbc.msn.com/id/28471626/ (accessed September 9, 2009).

11. Jonathan Marks, interview by the author, February 28, 2009.

12. Ibid.

13. Sharona Hoffman, "'Racially-Tailored' Medicine Unraveled," *American University Law Review* 55 (February 24, 2006): 395, 397–98.

14. Quoted in Emily Singer, "Beyond Race-Based Medicine," *Technology Review*, January 16, 2009, www.technologyreview.com/biomedicine/21972/ (accessed September 9, 2009).

15. Jerry Adler, "What's Race Got to Do with It?" *Newsweek,* January 12, 2009, www.newsweek.com/id/177737 (accessed September 9, 2009).

16. Ibid.

17. Peter Aldhous, "Genome Sequencing Falls to $5000," *New Scientist,* February 6, 2009, www.newscientist.com/article/dn16552-genome-sequencing-falls-to-5000.html (accessed September 9, 2009).

18. Mary Engel, "Cost of Genome Sequencing Falls," TampaBay.com, November 8, 2008, www.tampabay.com/incoming/article895229.ece (accessed September 9, 2009).

19. Ray Kurzweil, *The Singularity Is Near* (New York: Viking, 2005), p. 73.

20. Adler, "What's Race Got to Do with It?"

21. Troy Duster, Jay Kaufman, and Pilar Ossorio, "Ask the Experts Forum #2: Genetics, Race and Disease," from *Unnatural Causes...Is Inequality Making Us Sick?* PBS, 2008, www.unnatural causes.org/assets/uploads/file/Ask_the_Experts _Forum2.pdf (accessed September 9, 2009).

22. David R. Williams, interview by PBS, "David R. Williams Interview Transcript," from *Unnatural Causes...Is Inequality Making Us Sick?* 2008, p. 1, www .unnaturalcauses.org/assets/uploads/file/Interview-DWilliams.pdf (accessed September 9, 2009).

23. Ibid., p. 2.

24. Brooks and King, "Geneticizing Disease," pp. 7–8.

25. Williams, interview, p. 3.

26. Ibid.

27. Ibid., p. 6.

28. See Bill Cosby and Alvin Poussaint, *Come On People: On the Path from Victims to Victors* (Nashville, TN: Thomas Nelson, 2007), for more on this topic.

29. Brooks and King, "Geneticizing Disease," p. 6.

30. Duster, Kaufman, and Ossorio, "Ask the Experts Forum #2," p. 4.

31. "Backgrounders from the Unnatural Causes Health Equity Database," from *Unnatural Causes...Is Inequality Making Us Sick?* PBS, 2008, pp. 9–10, www.unnaturalcauses.org/assets/uploads/file/primers.pdf (accessed September 9, 2009).

32. Ibid., p. 10.

33. Ibid.

34. Brooks and King, "Geneticizing Disease," p. 6.

35. Warren E. Leary, "Social Links Are Seen in Black Stress" *New York Times,* February 6, 1991, http://nytimes.com/1991/02/06/us/social-links-are-seen-in -black-stress.html (accessed September 9, 2009).

36. Brooks and King, "Geneticizing Disease," p. 10.

37. US Census Bureau, "Household Income Rises, Poverty Rate Unchanged,

Number of Uninsured Down," *US Census Bureau News*, August 26, 2008, www
.census.gov/Press-Release/www/releases/archives/income_wealth/012528 .html
(accessed September 9, 2009).

38. Carmen DeNavas-Walt, Bernadette D. Proctor, and Jessica C. Smith, US
Census Bureau, "Income, Poverty, and Health Insurance Coverage in the United
States: 2007," p. 20, www.census.gov/prod/2008pubs/p60-235.pdf (accessed September 9, 2009).

39. American Lung Association, "Lung Disease Data at a Glance: Lung
Cancer," www.lungusa.org (accessed September 28, 2009).

40. American Cancer Society, *Cancer Facts and Figures for African Americans
2006–2007* (Atlanta, GA: American Cancer Society, 2007), p. 1.

41. Office of Minority Health, US Department of Health and Human Services, "Cancer and African Americans," June 26, 2009, www.omhrc.gov/
templates/content.aspx?ID=2826 (accessed September 9, 2009).

42. Steven H. Woolf, Robert E. Johnson, George E. Fryer Jr., George Rust,
and David Satcher, "The Health Impact of Resolving Racial Disparities: An
Analysis of US Mortality Data," *American Journal of Public Health* 94, no. 12
(December 2004): 2078–81, www.ajph.org/cgi/reprint/94/12/2078 (accessed
September 9, 2009).

43. American Sociological Association, "Race, Ethnicity and the Health of
Americans," July 2005, p. 11, www2.asanet.org/centennial/race_ethnicity_health.pdf
(accessed September 9, 2009).

44. Brooks and King, "Geneticizing Disease," p. 23.

45. Judith Lewis, "Moving Mountains," *Los Angeles Times*, January 6, 2008,
http://articles.latimes.com/2008/jan/06/books/bk-lewis6 (accessed September
9, 2009).

Chapter 6

RACE AND LOVE

Producing a mongrel child is the most serious physical sin one can commit.

—Racist Web site[1]

If any white person intermarry with a colored person, or any colored person intermarry with a white person, he shall be guilty of a felony and shall be punished by confinement in the penitentiary for not less than one nor more than five years.

—Virginia law, 1967[2]

Mating between members of different human groups tends to diminish differences between groups, and has played a very important role in human history. Wherever different human populations have come in contact, such matings have taken place. Obstacles to such interaction have been social and cultural, not biological.

—American Association of Physical Anthropologists[3]

Increasing rates of intermarriage with each generation and changing patterns of interracial marriage support the observation that we are in the midst of a quiet revolution.

—Maria P. P. Root[4]

Nothing quite derails the concept of race like a thing called love. Yes, despite the mighty and well-guarded walls of race that divide humankind, many people throughout history have managed to tunnel under, climb over, or burst through these walls in the name of love. Racists across the centuries have told us that different "kinds" of humans cannot or should not coexist, let alone love one another. Yet despite the claims of irreconcilable differences and conflicting biologies, people fall in love and have healthy, happy children. In their efforts to stop love from crossing the boundaries of race, people have made cruel laws, cited the demands of racist gods, and even committed murder. None of this worked, at least nowhere near to the degree that racists hoped for. Although interracial marriages are still a small minority of total marriages in the United States, they have grown by 500 percent since 1970.[5] In 1970, fewer than 2 percent of all American marriages were identified as interracial. Today more than 7 percent are interracial.[6] Not long ago, virtually every interracial couple was hit with this question: "What about the children?" It was a common belief that the offspring of interracial unions would be burdened by their ambiguous racial identity and would suffer for the "mistake" of their parents. Today, thanks to President Barack Obama, Tiger Woods, Halle Berry, Mariah Carey, and millions of other children who are doing just fine, I suspect this concern is not raised as often as it once was.

In general, American people have become much more tolerant with regard to racial issues in recent decades. According to the Pew Research Center, less than half of white Americans in the late 1980s agreed with the statement, "I think it's all right for blacks and whites to date each other." However, by 2003, 72 percent agreed with this statement. Acceptance is even higher among younger Americans. In 2002–2003, 89 percent of white people between the ages of eighteen and twenty-five agreed that it's okay for black people and white people to date each other.[7] An Asian-white, Hispanic-black, black-white, or any other combination was once an

unusual sight, guaranteed to turn heads in America and many other countries. Today, however, interracial couples are so common that they go
unnoticed in many cities and towns. The world has changed and the trans-
formation continues. But hold on: is this really anything new? According to
the anthropologists, whenever humans have come together throughout
history and prehistory, they have sometimes fought but always mated.
Members of our species have been hooking up across tribal, clan, religious,
national, and racial borders since … well, probably as long as there have
been such things as borders. Interracial love is an issue only because we
made it an issue by creating and believing in races.

FULL DISCLOSURE

It is a challenge for me to write about this topic, because I have lived it so
thoroughly. I have dated many women of many different "races." I am now
married to a Caribbean woman who has recent African, European, and
Asian ancestry. I have three children, whom many people would describe
as "mixed" or "multiracial." (I avoid those labels for them, because I am
aware that *everybody* is "mixed" or "multiracial.") It is difficult for me not to
roll my eyes and sigh as soon as the subject of interracial love comes up.
Why should I even dignify such a concept with comment or analysis? I am
so far down the road on this that it feels ridiculous to take seriously the
proposition that two people of different "races" are so radically different
that they shouldn't love one another. Sadly, however, I have no choice but
to take it seriously. The conflict and pain, both emotional and physical, that
have come from interracial love demand that we all care about the issue.

   In 1955, a young black kid from Chicago crossed a line while he was
visiting relatives in a small Mississippi town. Emmett Till, at age fourteen,
crossed that invisible line that is meant to keep people of different races
from loving one another. In some places, in some times, that line has been
deadly serious. No one knows precisely what happened when Till was in a
shop, but something got him into trouble. He may or may not have flirted
or whistled at a white woman who was in the store. He may or may not
have touched her and asked for a date. All that is known for certain is that
around midnight a week later, the woman, her husband, and two other men
drove to the house where Till was sleeping. They took him away to a

secluded area, beat him up, and then shot him. The killers tied a weight around the fourteen-year-old's body using barbed wire and then dumped him in the Tallahatchie River. The body was recovered three days later. Two white men, J. W. Milam and Roy Bryant, were arrested and charged with the kidnapping and murder. But they were acquitted after an all-white jury had deliberated for just sixty-seven minutes.[8]

Prior to the not-guilty verdict, Emmett Till's body had been taken back to Chicago for a funeral service there. His mother, Mamie Till, made the decision to have an open-casket viewing at Chicago's Roberts Temple Church of God. "I wanted the world to see what they did to my baby," she said. Thousands of people turned out to see the mutilated remains of the teenage boy. *Jet* magazine and the *Chicago Defender* published photographs of the corpse, and European newspapers covered the case as well.[9]

Younger readers may find it too bizarre or outrageous to accept that a fourteen-year-old teenager could be brutally tortured and murdered simply because he may have whistled at or flirted with a white woman, but it's true. According to statistics from the Tuskegee Institute archives, 4,743 people were lynched in the United States between the years of 1882 and 1968. Of these, 1,297 were white and 3,446 were black.[10] Numerous reasons were given for lynching black people. Not all black male victims were murdered because they had been accused of raping or flirting with a white woman, but many were. This reflected the level of concern that some white people had about interracial love. Race belief and the impulse toward the restriction of love do not have to lead to such barbarism, of course. But we know that they can.

These lynchings were horrific events that many times involved large crowds, including children, cheering and posing for photographs with the victims, who were often hung from trees in prominent places, perhaps as trophies or warnings. Of course, I would not go so far as to compare those who condemn "interracial marriage" today with those who tortured and murdered black men by the thousands in the previous two centuries. However, I would point out that holding such a position does seem to place one's mentality uncomfortably close to that of a lynch mob.

An American Dilemma: The Negro Problem and Modern Democracy, the landmark 1944 study by Gunnar Myrdal, surveyed the conditions black people faced in the United States. This work was influential in the Supreme Court school desegregation case, *Brown v. Board of Education of Topeka*, and many race-related government policies that came later. Myrdal included fascinating descriptions of the sexual taboos that were in effect through much of America but most severely in the South:

> The ban on intermarriage has the highest place in the white man's rank order of social segregation and discrimination. Sexual segregation is the most pervasive form of segregation and the concern over "race purity" is, in a sense, basic. No other way of crossing the color line is so attended by the emotion commonly associated with violating a social taboo as intermarriage and extra-marital relations between a Negro man and a white woman. No excuse for other forms of social segregation and discrimination is so potent as the one that sociable relations on an equal basis between members of the two races may possibly lead to intermarriage.
>
> Intermarriage is prohibited by law in all the Southern states, in all but five of the non-Southern states west of the Mississippi River. In practice, there is little intermarriage even where it is not prohibited, since the social isolation from the white world that the white partner must undergo is generally intolerable even to those few white people who have enough social contact and who are unprejudiced enough to consider marriage with Negroes.[11]

Activities such as eating, dancing, and swimming in pools or at the beach were all strictly segregated in the South, wrote Myrdal, primarily because white people believed there was a danger they could lead to interracial love and sex. Of course other issues, such as economic competition between poor whites and black laborers as well as a general uneasiness with regard to "different kinds of people," played into all of this as well. But it is interesting that fears of interracial love were so prominent in many aspects of daily life.[12]

So effective was the continual policing of the "color line" in the twentieth century that Myrdal noted: "even the old custom of white men keeping Negro concubines is disappearing in the South and is rare in the North."[10] Meanwhile, the black community in the South was not much more receptive to the idea of interracial love, in part because of "race pride." The black community was also likely to ostracize "mixed" couples.[13]

At risk of becoming too repetitive, I remind readers that races are not valid biological categories of human beings. It is important to keep this in mind, especially when reflecting upon a history in which people have been beaten, burned, dismembered, and hung from trees because they were accused of flirting with or kissing a member of another "race." Given the scientific facts of the last two hundred thousand years or so of humankind's existence, all human relationships qualify as "interracial." If we are all descended from the same small population of people living in Africa a relatively short time ago, then how can there be such a thing as pure races? Obviously, all problems with "interracial love" are cultural. People have always placed taboos on mating. Race is just another example.

LOVING

It's 2 a.m., July 11, 1958. All is quiet in the small town of Central Point, Virginia. Mildred and Richard Loving are sleeping peacefully in bed in their home. Just five weeks earlier, the two Americans married in Washington, DC. They couldn't marry in their home state, because interracial marriage was against the law in Virginia. The honeymoon ended, however, as they awoke to find the county sheriff and two of his deputies standing in their bedroom shining flashlights in their faces. "Who is this woman you are sleeping with?" barked one of the intruders.

"I'm his wife," said Mildred.

Richard pointed to the marriage certificate he had hung on the wall.

"That's no good here." said the sheriff.

Yes, hard as it may be for many people to believe today, Mildred and Richard Loving had committed a crime simply by being married and living in Virginia. At that time not only was it against the law for an interracial couple to get married in Virginia, but it was illegal for them to live there as a married couple even if they had been legally married in another jurisdiction. I am sure this sounds unbelievable to many Americans today, especially young people, but it was the harsh reality faced by those who were in interracial relationships in many places not so long ago. In all, eighteen American states (more than half) had antimiscegenation or anti–race mixing laws at one time or another.[14]

The Lovings faced a year in jail for violating the Racial Integrity Act but were given the option of a plea bargain. If they would agree to leave the state and not return, as a couple or apart, at the same time for twenty-five years, they could go "free." They took the deal and left Virginia. While it may have been better than doing a year in jail, this still was not a good result for the couple. It meant that Mildred and Richard Loving were criminals, felons with records. They had dared to love and marry across a culturally created border between culturally created races and were exiled for it.[15]

This punishment given by the Virginia legal system for breaking what today sounds like some prehistoric tribal taboo seems preposterous by today's standards. Banished for twenty-five years because you fell in love and got married? Banned from your home state because your spouse is not of the same race? To make matters worse, the Virginia circuit court judge who ruled on the Loving case, Leon Bazile, was clearly biased by his religious beliefs. "Almighty God created the races white, black, yellow, Malay and red, and he placed them on separate continents," declared the judge in his opinion. "And but for the interference with [God's] arrangement there would be no cause for such marriages. The fact that he separated the races shows that he did not intend for the races to mix."[16]

The Lovings moved to Washington, DC, where they lived for a few years. Unhappy and frustrated that they could not live in the only place they had ever considered home, however, the Lovings decided to fight back. Mildred, aware of the growing Civil Rights Movement, wrote to Attorney General Robert Kennedy in 1963 and described their situation. Kennedy put the Lovings in touch with the American Civil Liberties Union (ACLU).[17]

ACLU lawyers Bernard S. Cohen and Philip J. Hirschkop asked Judge Bazile to overturn the verdict in the Loving case. He did not, of course, so Cohen and Hirschkop appealed to the Virginia Supreme Court of Appeals. That court also stuck by the original verdict, so the case ended up in the highest court in the land, the United States Supreme Court. While it might be tempting to think of Mildred and Richard Loving as larger-than-life superheroes, champions of justice who wanted to make history, that just wouldn't be accurate. They showed courage, yes. However, by all accounts, they were just good people who loved each other and wanted nothing more than the right to live together and raise their family like any other couple in America. "We have thought about other people," Richard said just before the US Supreme Court reached its decision, "but we are not doing it just because somebody had to do it and we wanted to be the ones. We are doing it for us—because we want to live [in Virginia]."[18] Richard asked one of the lawyers to tell the court just that: "I love my wife and it's not fair that I can't live with her in Virginia."[19]

"They were very simple people, who were not interested in winning any civil rights principle," the Lovings' lawyer, Bernard Cohen, said in a 2007 National Public Radio interview. "They just were in love with one another and wanted the right to live together as husband and wife in Virginia, without any interference from officialdom. When I told Richard that this case was, in all likelihood, going to go to the Supreme Court of the United States, he became wide-eyed and his jaw dropped."[20]

To understand the absurdity of the Lovings' "crime," one only has to read from Section 20-54 of the Virginia Code that they were convicted of violating: "It shall hereafter be unlawful for any white person in this State to marry any save a white person, or a person with no other admixture of blood than white and American Indian. For the purpose of this chapter, the term 'white person' shall apply only to such person as has no trace whatever of any blood other than Caucasian; but persons who have one-sixteenth or less of the blood of the American Indian and have no other non-Caucasic blood shall be deemed to be white persons. All laws heretofore passed and now in effect regarding the intermarriage of white and colored persons shall apply to marriages prohibited by this chapter."[21]

"No trace whatever of any blood other than Caucasian"? Sorry, Virginia, that ship sailed long before the Lovings came along. But this fantasy of racial purity is always at the heart of objections to interracial love. The matter is not so much one of tolerance for others, change, or liberal progress as it is of simply recognizing reality. More than anything, opposition to interracial love contradicts the true state of America's gene pool and the world's gene pool. People are not wrong to oppose interracial marriage only because it denies people basic rights and freedoms, but they might also acknowledge that the entire premise is based on a lie. Pure races don't need protecting, because pure races do not exist. But the need to preserve the integrity of white blood was and is seen by many as nothing less than critical to the fate of civilization. Consider these portentous words from a former Mississippi governor and senator:

Once the blood is corrupted, there is no power on earth, neither armed might, nor wealth, nor science, nor religion itself, that can restore its purity. Then there will be no Negro problem because the blood of that race will be commingled with the blood of the white race, and a mongrel America would have no reason to worry over the race issue. Who will choose that our free Republic of tomorrow be the product of miscegenation, bastardization, and mongrelization? We should be eternally grateful that there is yet time for the other choice. Separation of the races is the only way to solve properly, adequately, and permanently the race problem and safeguard the future of this Republic. No obstacles are insurmountable when the life's blood is at stake. The blood, culture, and civilization of the white race are our heritage.[22]

Those charming sentiments are from Governor Theodore G. Bilbo's book, *Take Your Choice: Separation or Mongrelization*, published in 1947. Don't think such ideas are extinct. The World Wide Web today has many sites that promote this same theme of racial purity while warning of the "dangers" of interracial love.

The Lovings may have been going up against centuries of tradition, racism, and entrenched ignorance in 1967, but their lawyers had a strong case and were more than capable of arguing it before America's highest court.

"The State is ignoring a very important point which we cannot overemphasize," Cohen told the Supreme Court justices, " and that is the right of Richard and Mildred Loving to wake up in the morning or to go to sleep at night knowing that the sheriff will not be knocking on their door or shining a light in their face in the privacy of their bedroom for illicit cohabitation.... The Lovings have the right to go to sleep at night knowing that if they should they not wake in the morning, their children would have the right to inherit from them. They have the right to be secure in knowing that, if they go to sleep and do not wake in the morning, that one of them, a survivor of them, has the right to Social Security benefits. All of these are denied to them, and they will not be denied to them if the whole antimiscegenism scheme of Virginia [is] found unconstitutional."[23]

It was 1967, thirteen years after the landmark Supreme Court case, *Brown v. Board of Education of Topeka*, which ruled America's public schools could not be segregated. It was three years after the passing of the Civil Rights Act, which gave all Americans the right to free access to hotels, restaurants, theaters, and so on. Fortunately, the momentum of social progress was in the Lovings's favor. As rare and as unpopular as interracial marriage may have been in the United States in 1967, the Lovings were on the right side of history.

The US Supreme Court delivered a unanimous decision on June 12, 1967. "There can be no question but that Virginia's miscegenation statutes rest solely upon distinctions drawn according to race," wrote Chief Justice Earl Warren:

> The statutes proscribe generally accepted conduct if engaged in by members of different races. Over the years, this Court has consistently repudiated "[d]istinctions between citizens solely because of their ancestry" as being "odious to a free people whose institutions are founded upon the doctrine of equality."
>
> Marriage is one of the "basic civil rights of man," fundamental to our very existence and survival.... To deny this fundamental freedom on so unsupportable a basis as the racial classifications embodied in these statutes, classifications so directly subversive of the principle of equality at the heart of the Fourteenth Amendment, is surely to deprive all the State's citizens of liberty without due process of law. The Fourteenth Amendment requires that the freedom of choice to marry not be restricted by invidious racial discriminations. Under our Constitution, the freedom to marry, or not marry, a person of another race resides with the individual and cannot be infringed by the State.[24]

"We were so very, very happy," recalled Mildred Loving. "I can't describe the way I felt. It was as if I'd been freed to live my life."[25]

The Lovings and their three children seemed to have finally settled down to lead a normal life in Virginia. Sadly, however, they would not grow old together. Richard was killed in an automobile accident in 1975. Mildred died in 2008 at the age of sixty-eight. She was survived by eight grandchildren and eleven great-grandchildren.[26]

The Lovings may not have sought lasting fame, but they got it nonetheless. Ken Tanabe, a New York City graphic design artist, is the product of interracial love and views the Loving case as critical to his very existence. Without it, he says, he might not have been born, because his parents probably wouldn't have gotten married. Tanabe was so moved by the profound implications of the Loving case and what he sees as the need for more attention to the growing and "positive" phenomenon of children from interracial unions that he started an annual day of celebration called

"Loving Day." It's celebrated on or around June 12 each year. Tanabe and the members of his movement hope that Loving Day will one day earn official status as a national holiday honoring the Lovings and celebrating the acceptance of interracial love.

"I discovered *Loving v. Virginia* completely by accident," Tanabe said. "I realized that this Supreme Court case was a milestone in the history of civil rights. I also realized that the case was virtually unknown among [those of] my generation. I felt strongly that everyone should know that interracial couples were once illegal, and that that discrimination was tied to slavery, segregation, and other institutionalized racism. I still feel that way. Education is one of the best tools to fight prejudice."[27]

Tanabe's optimism has energized Loving Day. It's about a historical event, yes, but it's also about building a more sensible world through education and awareness. And everyone is invited, not just the "mixed" crowd. The first Loving Day was celebrated in 2004. Since then the movement has grown, attracting media attention from around the world with events held in New York City, Los Angeles, and other cities.

> When I see people of every description coming together for Loving Day, it does give me hope for the future—but not in the way one might expect. Loving Day is an annual tradition to be shared among friends and between generations. It fights racial prejudice through education and it builds multicultural community. Loving Day creates a positive environment for multicultural couples and individuals, but it's really for anyone who shares our mission and values. In fact, we specifically welcome people who don't identify as multicultural. I've heard people say that interracial relationships and people are "the future" and they are the solution to racism. In fact, racism is totally separate. Your "race," or your partner's race, doesn't necessarily make you any more or less racist. You didn't have to be black to march with Dr. King, and you don't need any ethnic "credentials" to be a part of Loving Day. The future is about educating ourselves as a society, fighting racial prejudice, and building multicultural community and identity.[28]

As a child, Tanabe says he experienced "playground-variety taunting" about his Belgian and Japanese ancestry, but he's not complaining. "I feel that the idea of being somehow 'diminished' through multiethnicity plays into old 'tragic mulatto' stereotypes. My multiethnic heritage inspired me to create Loving Day. So for me, the positive experiences that come from my identity have been far more significant than the negative. I believe that broad 'racial' classifications are of limited use. They're inaccurate and they create more problems than they solve. I believe that we can move past old categories and divisions. Loving Day is a great way for us to work together toward that goal."[29]

THE IN-BETWEEN PEOPLE

There is no way to know such a thing, of course, but what if Bob Marley had inherited all of his musical talent from his white father? Would that have made "Redemption Song" less soulful or "Concrete Jungle" less haunting? And what if Malcolm X's white ancestry had been the source of his talent for oratory? Would that have made his fiery words about justice less sincere? Some people place so much emphasis on their concept of race that one wonders if they sometimes lose sight of the person, the individual who is standing there behind that thick cloud of race belief. What is it like to be a "mixed" or "multiracial" person? Are such people more, or are they less, than those who are fully of one race? How does it feel to have different kinds of blood coursing through your veins? If you have ever wondered about this, well, simply ask yourself, because you are a "mixed" or "multiracial" person. Everyone is.

This was never more clear to me than a few years ago, when my son and I participated in *National Geographic's* Genographic Project. The aim of the project is to collect DNA samples from hundreds of thousands of people, in order to produce a genetic atlas of humankind that will show in new detail how we spread around the world from Africa. Although it didn't surprise me, it was an exciting moment to see the world map with my ancestral trail that the Genographic Project sent me. The personal history book that is written into my DNA revealed the journey my ancestors, my genes, had taken over the last 50,000 years or so. I love following that migration line with my eyes. It starts in Africa (I am African!), then pushes up into the fringes of Asia (I am Asian!). Finally, it veers over into Europe (I am European!). It was exactly what I had expected, yet I was still moved by the sight of that map. For me, that path, derived from a DNA sample scraped from the inside of my mouth, represents life, struggle, and triumph. My people—those glorious Africans who came up with ingenious ways to overcome the dangers of their world—feel close when I think about my past. They outsmarted predators. Their powerful minds found new ways to improve their lives and increase their odds of survival. My Eurasian kin faced challenges such as severe cold and perhaps even encounters with *Homo erectus* and Neanderthal populations. Many humans died too young, but none of my direct ancestors did, at least not until they had passed on their genes to the next generation. They all survived everything the world threw at them long enough to build upon the road to me, here and now—just as your ancestors did for you. In light of this, to think of myself as a member of one sliver of humanity called a "race" or as nothing more than a citizen of a nation would be shamefully incomplete and inaccurate. I am so much more than those things, as are all people. The world may call me white, but I know what I really am. I'm mixed, multiracial to the bone. I may not be a member of a pure race, but I'm pure human and that's better. To think of myself as anything less would be disrespectful and dismissive to the people who pushed my genes forward across all those continents and all those thousands of years. Their strength and intelligence are the reasons why I'm alive.

What the reality of our origins and migrations says to those who would worry about the "unnatural" condition of interracial love is that there could be nothing more natural than interracial love. What is unnatural is refusing to embrace all humankind as our extended family and resisting the reality that all potential mates are potential mates. As for the idea that "mongrel" offspring are somehow bad, nothing could be more ridiculous. "Mongrel" sounds like such a bad word, but if it means the product of something other than a pure race, then we are all mongrels and the word loses the power to insult.

The American Academy of Child and Adolescent Psychiatry says that "multiracial" children may face some unique challenges but they tend to be high achievers with a strong sense of self, regardless. The organization has produced an interesting fact sheet about these children. Some of the items on it are as follows:

- About two million American children have parents of different races;
- In the United States there has been a 1,000 percent increase in marriages between whites and Asians;
- In a recent survey, 47 percent of white teens, 60 percent of black teens, and 90 percent of Hispanic teens said they had dated someone of another race;
- Recent research has shown that multiracial children do not differ from other children in self-esteem, comfort with themselves, or number of psychiatric problems. Also, they tend to be high achievers with a strong sense of self and tolerance of diversity;
- Children in a multiracial family may have different racial identities from one another. Their racial identity is influenced by their individual physical features, family attachments and support, and experiences with racial groups;
- To cope with society biases, mixed-race children may develop a public identity with the "minority" race, while maintaining a private interracial identity with family and friends;
- Research has shown that children with a true multiracial or multicultural identity generally grow up to be happier than multiracial children who grow up with a "single-race" identity.[30]

"Mixed-race" people have always been present in America, of course, but the United States government only began recognizing them in the year 2000. In the census of that year, Americans had the option of choosing "some other race" or selecting a combination of races. That resulted in a count of 6.8 million "multiracial" Americans. But that total is considered by many to be inaccurate, due to the way the options were set up. The government's annual Population Estimate Program allowed Americans to claim to be of "two or more races" and is thought to have produced a more realistic number. In 2008, the program revealed that 4,856,136 Americans claim to be of "of two or more races."[31] A total of fewer than five million "multiracial" Americans may seem like a small number for a country of more than three hundred million. But it is a number that is growing fast. The US Census Bureau projects that the number of people who identify themselves as being of "two or more races" will more than triple to 16.2 million by mid-century.[32]

I stress the need to keep these statistics in their proper context. The greatest change in America has been the legalization of interracial marriages, the reduction of social disapproval of them, and the willingness of more people to openly acknowledge their ancestry. Never forget, however, that Americans have been making "interracial" babies since before the United States came into being. The significant difference between the past and the present is that in most cases in the past, the children would choose one race or have one race imposed on them. Yes, there was some recognition of "multiracial" people in some regions during America's past: "mulattoes," "quadroons," "octoroons," "redbones," "mestizos," and so on. But generally, society's unease over "race mixing" meant that one race and only one race must be selected or designated.

That number of "multiracial" Americans is misleading for another reason. It reflects a racial scorecard that was created by culture but is viewed as purely a biological matter by most people. Genetically speaking, however, the popular concept of "multiracial-ness" makes no sense whatsoever. Most black Americans have "white" genes (meaning genes from people who are or are descended from "white people" who lived in Europe for some extended time in the last twenty thousand years or so). Every white person in America has African ancestors. Maybe I'm going too far with this, but when a black person and a white person marry in America, isn't it really just a case of two Africans getting back together? One may have many ancestors who were in Africa a few centuries ago and the other has many ancestors who were in Africa sixty thousand years ago. It's just a difference in time: what's the big deal?

IT'S LIKE RACES AREN'T EVEN REAL

People of virtually every race are mixing these days, just as they always have done. The past, present, and future are clear. America and the world are blending. All this free-flowing love leads one to suspect that racial divisions don't matter nearly as much as people have been led to believe over the last couple of centuries. The larger and more important question here is about what the rise of interracial marriage and "multiracial" children says about the meaning of race in the twenty-first century. By openly and legally acknowledging what has always been there, societies such as the

United States are dismantling the concept of race, whether this is recognized or not. During President Obama's election campaign, for example, I noticed that some people raised the question of why he is black if his mother was white. The one-drop rule, the idea that white blood is contaminated by black blood and a person of black and white ancestry is therefore "black," doesn't seem to be going over as well as it did in previous centuries. People appear to be thinking more and asking more questions about race. This can only be a good thing. The more people listen to their curiosity about race and challenge convention and tradition, the more likely they are to discover that the concept of race is a sham.

SO, WHAT'S IT LIKE BEING MARRIED TO AN ALIEN?

It might disappoint more extreme race believers to learn that interracial couples tend to have relationships that are remarkably similar to other relationships. It is unlikely, for example, that most "mixed" couples fixate on their racial differences. If they did, I would suspect they were heading for troubles down the road. An interracial marriage is first and foremost a marriage. This means that things like leaving the toilet seat up, having similar tastes in television viewing, and tolerating mothers-in-law end up being far more important than minute variants in prehistoric ancestry. Whether interracial couples are Hispanics, Asians, Native Americans, white, black—whoever they are—interracial love is likely to come with all the joys, challenges, and agonies of any other pairing. Yes, racist strangers or family members may place additional pressures on a couple, with cold stares, ostracization, or even threats of violence. Even then, however, love is the key.

"Ultimately, it's simple: You love who you love, if you're lucky enough to find love," writes Karyn Langhorne Folan, an American novelist who is black and married to a white man. "As Mildred Loving knew, it's not about black or white at all. Her case and our Loving Day celebrations have little to do with love or life. Making a marriage work lies in all the things that make up life as a man and woman: paying bills, raising children, trying to keep the connection between you vibrant and alive. Color isn't a part of that."[33]

It is difficult for me to judge whether I am the best person or the worst person to write this chapter. I am either positioned perfectly or positioned far too close to see interracial relationships with the necessary objectivity. I have more experience at this thing called interracial relationships than most people—considerably more. If I could gather every woman I have ever dated into one place at one time, it would look like a meeting of the United Nations General Assembly. Before getting married, I was fortunate to feel free enough to live and love with an openness that was totally unbounded by racial nonsense. I have done things, seen things, traveled beyond where tourist busses go, and loved beyond where race believers go. All in large part because I did not allow an unscientific and illogical belief in biological races to trap me inside a myopic or stunted worldview.

I always have been attracted to women based on their beauty, personalities, and brains. To my great benefit, racial affiliations were never deal killers for me. If a woman was interesting to me, I saw no reason not to politely introduce myself and ask for her phone number or permission to enter her village. I could not have cared less about genetic lineage or racial affiliation. Never once did I date a woman who seemed like an alien from outer space. (Well, okay, there were a few, but they were just eccentric. It had nothing to do with race.) Never once did I feel there was a distance between us that could not be fully explained by culture or personality. *Cultural* race may have generated the occasional gap, but *biological* race never did. My conclusion is that the variation of women *within* races is far greater than the variation of women *between* races. I suppose the notion of biological race could have been a problem if I had believed in it and thought about it constantly. In my view, when it comes to love, once the lights dim and the candles are lit, only the saddest of fools waste time thinking about melanin content and cranial contours.

"I LOVE BEIGE CHILDREN"

Let's consider the thoughts of people around the world who are in, or have been in, "interracial relationships." Do they think wading into "interracial love" is like exploring a distant galaxy? Or do they see it as a little less dramatic than that? I asked several people how they felt about their "interracial" experiences and how they feel about the concept of loving across the borders of race in general. The following comments are drawn from my interviews with them.

Satina, thirty-three, "brown-skinned," self-employed Caymanian, is married to a "white" Englishman and reports that her love life has always been determined by factors such as culture and personality. "Race has not been a factor," she says.

Retired "white" British police officer Mike, 44, has been married to a "black" woman for fifteen years. "Still going strong," he says. "An interracial relationship to me is like any other relationship I've experienced. Race is not an important factor or not even a factor at all."

Kimberly, thirty-six, is a "black" woman in New York City who loves the "confusion" that often comes with interracial relationships. Now married to a black American man, she has been in interracial relationships in the past. "The novelty was more significant as a teenager—it was less common than it is now. I love interracial couples; I love biracial kids. I love challenging old notions and stereotypes. I love it when a black person looks white and when a white person sounds black. I love the confusion across faces when someone sees a black mother with a white child. I love beige children—especially those where it's difficult to determine by looking what they *are*."

Joshua Lipana, seventeen, is a student in the Philippines who describes himself as a mix of Filipino and Spanish. He says the interracial relationship he had was "pretty cool, nothing to take note of."

Tina, a Caribbean woman married to a white Englishman, once had a confrontation with a black security guard at a museum in the United States. "He gave off some pretty heavy bad vibes to us and when he finally could not help himself he came over to me and confronted me," she says. "He asked if I did not feel bad about betraying black people by dating someone white and what could I possibly have in common with him. He said the usual stuff about slavery and what white people did to the blacks. Needless to say, I had a few words to say back to him."

Tina says the few disagreements between her and her husband have involved minor differences of opinion about things such as how they should address their elders. "But I consider things like that a cultural rather than race issue," she says.

Tom McCallum, forty-three, is a "white" Scottish company director who has been married to a Caribbean woman for sixteen years and says race has had "no impact" on their relationship. He adds that their children have never paid much attention, if any, to race. McCallum feels the world should just abandon the concept of race altogether. "Our kids already have," he says. "It's just not on their radar."

Susan, thirty-seven, is an entrepreneur and dance instructor who doesn't consider herself to be a member of a race. "If I'm forced to answer I say 'mixed.' I figure, what the hell, keep them guessing." She sees no place for race in love. "For me, race is not an important factor, otherwise the relationship cannot exist. The individuals in an interracial relationship have to see beyond race in order to come together as a couple. Race becomes a factor when others are brought into the relationship, like friends and family members, and when social pressures are applied as others try to define the relationship. The effect that that pressure will have on the couple will depend on the importance they place on it."

John, fifty-nine, is a "white" Englishman who has had several interracial relationships. Although he is disturbingly pessimistic about the state of racism around the world today, as well as about prospects for the future of all civilization, he says a romantic connection with someone else, if sincere, can transcend race. "When you get on with people, race is totally irrelevant," he says.

Kevin, an American graphic artist, says he is "white, if you believe what my driver's license tells you." He considers himself to be a mix of Irish, Scottish, Polish, and American Indian. He has been in several interracial relationships, which "varied from exciting to tragic," just like his other relationships, he says. Race was never a factor in the success or failure of the relationships, he explained.

Carlene, a forty-seven-year-old property manager in the Cayman Islands, says she doesn't think about race much, but if she had to identify herself that way it would be "Negro," She has been married to two "white men." (No, not at the same time.) Their different races meant nothing living in the Cayman Islands, she says. However, when they traveled in the United States, she "definitely" noticed a difference in the way people looked at them and acted toward them.

Graphic design artist Camille says she is a "mutt," a "little bit of everything":

> I'm mixed, my parents are mixed, and my grandparents are mixed. Seriously, I don't think about me having a race. If you are asking me if I'm black or white, I don't see myself as either. I suppose people would round

me off as black. But if you want to get specific with colors, I usually tell people I'm honey-coated. My skin color doesn't make me who I am.

I've dated different nationalities with various skin colors. For me, I don't care what color you are as long as there is a mental and physical connection between [us]. If [people] were to limit their options to dating only a certain race, then they would be cheating themselves out of the possibility of finding a perfect match. I need a certain kind of mentality, and if that mentality that I crave happens to come wrapped in a sexy white, black, brown, or purple package with a cute smile... hey, it's like Christmas every day for me.

Wendy, a "white" journalist from the UK, has lived and worked in Africa and the Caribbean. She has had several interracial relationships and says race was never an issue. "The men I have dated have come from a similar cultural perspective and educational background. It's fair to say that when I was in Kenya, it would have been very unlikely that I would have dated a Maasai warrior, for example. As hot as the dude may have been, we wouldn't have had much common ground—and that is the issue."

Christopher Olson, a "white" twenty-three-year-old Canadian university student, has never experienced an interracial relationship. But don't think he's not eager to give one a try, the first chance he gets. "I've never been in an interracial relationship before, but I've *wanted* to get together with women of other races," he said. "It was always a coincidental occurrence if someone I found attractive was of another race. Frankly, a woman's hair color is probably a more determining factor in whether I want to sleep with her than whether her skin is [white] or barn-shed brown."

Kourtni, a school administrator who has lived in both the United States and the Caribbean, has a "white" mother and a "black" father. If pressed, she describes herself as "mixed." "I've dated guys of different races, and I don't feel that there was a difference between any of them [based on race]. To me race just wasn't a factor."

Robert, thirty-nine, is a "white" American journalist with no inhibitions about dating nonwhite women. He views all relationships as difficult, so the interracial aspect is just one more thing to deal with. "You simply never know if her relatives, [your relatives], outsiders, friends, colleagues or co-workers don't approve and will make that known. Just one more thing to deal with, I guess. I've been hurt seeing racism affect someone I'm dating," he says. "In restaurants or other public places, I've seen people look down on someone I've brought there on a date, simply because we're a mixed couple. Hard to articulate, but the scorn isn't directed at me— from people of my own race or another— but at the female. Even those of her own race cast sideways glances or otherwise make it known that they don't approve of her or us or whatever it is that's bugging them."

Robert is now in the American South, but he lived in the Cayman Islands several years ago. The diversity of the population and the absence of rigid race-based social barriers impressed him. "Eventually, we'll all move beyond racism, because there are so many mixed marriages and multicultural children that hopefully it'll end up like the Cayman Islands. Can't hate someone else if you're not even sure of your own lineage. Even the short time I lived in Cayman, it was refreshing to see very little racial discrimination. Now, economic discrimination—that's another story."

She doesn't want her name used, but the comments of one particular "black" woman (with one "white" grandparent) are just too good to leave out. She says the idea of interracial relationships was an issue when she was younger but not now. One certainly hopes it's not an issue, because today she is the mother in what might be described as the Brady Bunch of the twenty-first century. She was married to a black man and had two children. They divorced. Now she is married to a white man who has three children from a previous marriage to a white woman. The result is a family that seems like a tiny melting pot all of its own. So far, so good, the mother says.

"I don't often think of my children as 'white' or 'black' kids, but just as kids," she explains:

> However, I do think it might be perceived as harder or weirder at times for a black woman to be the mother of white kids than vice versa. One of the things I have discussed with my white children is the fact that it is sometimes difficult for them because there is an immediate assumption based on visual evidence that I am not their mother and so we can't 'blend' as a family in that way. Having both black and white children also leads to the challenge of balancing their experiences of the world and wondering whether they will have different experiences based on race— in the same way that I imagine a mother of disabled and able-bodied children might contemplate this. It has never once been an issue that has been raised expressly with me by anyone other than my children, and no one seems to react visibly to our racially mixed family in any way other

than with delight. I suppose we have been lucky, but I think that is also a sign of how times have changed, at least for us.

But don't think for a second she believes racism is a thing of the past. She says her own encounters with racism alone could fill a book:

Name calling as a child, being underestimated as an adult, and constantly having to prove that I am as good as my peers professionally. Being pulled over by police when driving in a fancy car with black male friends. Having my brothers experience the same police harassment. Racist taunts shouted at me and my children on the street in London—this was in the '90s, so I am not sure this would happen quite so often now. Having to explain racism to my children or having to witness my then six-year-old being called a racist name by a woman in Scotland. Watching young black girls trying to measure up to standards of beauty for hair and skin that are set without taking them into account—and being one myself.

I am not sure that racism will ever disappear completely, because it seems to be a human instinct to attempt to assert our own superiority on the basis of perceived distinctions of class, race, wealth, beauty or some other standard. The other side of that coin is the misguided attempt to find strength or validation in creating a pack [that] is formed against outsiders and living by the rules of that pack to the exclusion of others. However, I think it is possible that racism will become largely unacceptable, so that people with racist views will find themselves in the minority rather than the majority. I also believe that this will diminish our experience of racism but that there will always be attempts at differentiation and subjugation of "others," as in sexual orientation, religion, [and so on].

German hairstylist Birgit says she would describe herself as "Caucasian" or "whitey" but says race has no meaning in her life at all. Her interracial relationships have not been very different than her other relationships, she says. "Maybe you have to adjust the menu a little bit, but other than that it feels the same to me. It's all about love and understanding and compromising." She adds that kids from interracial relationships are "usually beyond beautiful! It's all in the mix!"

Tauriq Moosa, a twenty-two-year-old "Indian" South African, says all of his relationships with women have been interracial:

But, it appears to me that no one has been in anything *but* an interracial relationship, since we can't all be 'one' race, because it does not mean anything and how are you meant to know whether you are truly white? For example, a white boy and a white girl are going out. It appears to be a non-interracial relationship. But what happens if the girl's mother is black and her father white? What happens if the boy's great-grandmother was Indian? That seems to me to be an 'interracial' relationship. Anything else is simply an arbitrary distinction based on how good your vision is, to distinguish hues or match them. That is, the only way that anyone could be said to be in a relationship that is not interracial is to have them stand side by side, with a Dulux color palette to compare them. If their exact color matches, then you have a non-interracial relationship. If it seems utterly bizarre, then you understand how I feel about considering any relationships non-interracial."

Retail manager Tamika, twenty-five, chooses to be flexible with her identity. "I would have to take a guess and say society sees me as mostly white," she explains:

As for myself, it [is] where I am and who I am with that determines whether I identify with my white heritage or my black heritage. On immigration forms, I tend to always check 'other.' I don't have to look at myself all day, so I don't seem to notice what color I am. It's like a person with a tattoo on their back; they tend to forget they even have it until they see it in the mirror or someone points it out. For a moment you are aware of it, but then life goes on and you forget all over again. I would have to say my mother set a good example in this department. I guess just by the way that she lived her life, I could see that it didn't matter what color you were or where you were from.

"I guess you would have to say that all of my relationships have been interracial, since I don't identify with one group over another," she adds. "Relationships have their ups and downs, so having dated from two different groups [black and white], I would say that there is no difference. Race is only a factor if you or the people around you make it one. Being a product of an interracial relationship, I see nothing wrong with being a bi-racial child. I would have to say it may have its benefits; you don't really belong to one group or the other, so you get to choose which group you want to belong to, and if you are spoiled like me, you choose both."

THE QUIET REVOLUTION

Clinical psychiatrist Maria P. P. Root, author of *Love's Revolution: Interracial Marriage*, calls those who are in interracial relationships "scouts on the front lines of a quiet revolution."[34] Root interviewed more than 175 families and discovered many interesting details. For example, she found that many young people today who are in interracial relationships that are not black-white do not even think of themselves as being in interracial relationships. It says a lot about the nature of race belief that recognition is influenced by the degree of troubled history between two particular races. Root points out the important truth that intermarriage has been going on around the world "since time immemorial."[35] She also warns against seeing interracial relationships as the perfect solution to racism. "We are headed toward an intimately interracial and multiracial country," she writes. "But make no mistake: Intermarriage does not mean the end of racial prejudice or bigotry—not necessarily even within those marriages. Interracial marriage is not this country's solution to its long-lived, seemingly intractable racial problems, but it is one indicator that race relations are changing."[36]

Root is correct that interracial love is not the only answer or solution to racism. It is, however, one powerful response to racism nonetheless.

She sates, "While hate, fear and anger are common responses to interracial marriage and have powerful short-term effects, love has proved a formidable opponent. With stealth, persistence, and a few legal twists, love carves a more hopeful path for future race relations through the sacrifices of the many who have made and make their commitments to love for better or worse."[37] In response to those who respond negatively to interracial marriage, Root has summarized much of her research in "Ten Truths about Interracial Marriage."

TEN TRUTHS ABOUT INTERRACIAL MARRIAGE[38]

1. The civil rights movement [in the United States] and subsequent patterns of racial desegregation created opportunities for people to interact in meaningful ways, which has resulted in an increased rate of interracial marriage.

2. In the past twenty-five years, women's decreased financial dependence on their families has given them freedom to choose mates regardless of family approval.

3. Love, shared vision, and common values compel an interracial couple to marry, just as they do other couples.

4. The motives behind interracial marriage seldom include the desire to rebel or to make a political statement.

5. Families that reject an interracial marriage value the reproduction of their race over love, integrity, and commitment.

6. In order to live in an affirming emotional climate, an interracial couple may have to replace estranged blood kin with fictive family friends.

7. Conflicts within interracial marriages are more likely to arise from cultural, gender, class, social, and personal differences than from racial ones.

8. Irreconcilable differences within interracial marriages are similar to those within same-race marriages: loss of respect, unwillingness to compromise, hurtful actions, lack of responsibility, dishonesty, and conflicting values.

9. The rate of divorce for interracial marriages is only slightly higher than for same-race couples in the continental United States; the gap is quickly closing as divorce rates rise for all marriages.

10. Interracial couples can and do produce healthy, well-adjusted children.

Root also wrote the following "Bill of Rights for People of Mixed Heritage" for the millions of children who come from interracial relationships:

BILL OF RIGHTS FOR PEOPLE OF MIXED HERITAGE[39]

I HAVE THE RIGHT...
> Not to justify my existence in this world.
> Not to keep the races separate within me.
> Not to justify my ethnic legitimacy.
> Not to be responsible for people's discomfort with my physical or
> ethnic ambiguity.

I HAVE THE RIGHT...
> To identify myself differently than strangers expect me to identify.
> To identify myself differently than how my parents identify me.
> To identify myself differently than my brothers and sisters.
> To identify myself differently in different situations.

I HAVE THE RIGHT...
> To create a vocabulary to communicate about being multiracial or
> multiethnic.
> To change my identity over my lifetime—and more than once.
> To have loyalties and identification with more than one group of
> people.
> To freely choose whom I befriend and love.
>
> © Maria P. P. Root, PhD, 1993, 1994

WAITING ON THE INEVITABLE

In hindsight, I am a bit surprised by how easy it was for me to transcend
and forget racial boundaries when it came to girls and, later, women. I
never pondered the implications of crossing a sacred border or violating
the implied lessons of the "Curse of Ham," as explained to me by some
Christians. I never worried that any "mixed" children I might create would
be tragic mutants condemned to live on the outskirts of a society that had
no place for them. I had more faith in America and the world than that. I
never felt guilty about betraying my race because I felt I was being loyal to

something greater—all humanity. I have never lost a wink of sleep second-guessing my decisions on those I have chosen to date and marry.

Time has proven me right for not believing in the imaginary walls of race. After twelve years of marriage, my wife and I have never had a single argument or disagreement that had anything even vaguely to do with race. I even asked her if I had forgotten something, but she confirmed it. We are not aware of any distance between us due to biological race or even cultural race. I may feel strange every time I walk out of my front door into this odd world of ours. But I never feel strange in the presence of the woman I love. Maybe it helps that we live in the Cayman Islands, a society that has so many "interracial marriages," or at least racially ambiguous pairings, that any other pairings seem to be in the minority. (I can't know for sure, however, because no such statistics are kept. In the Cayman Islands neither marriage licenses nor birth certificates ask for racial identity.) Granted, external problems may be imposed on a couple by society or by extended family members. From our perspective, however, it seems that race is a potential problem for those in interracial relationships only if one or both partners believe in it and make a big deal about it. Funny, isn't it? The concept of race loses so much power when you stop believing in it. Maybe that's because it was never really there in the first place.

I remember having a discussion about racism with a college-educated black woman in the American South. I asked her if she would ever consider dating a white man. "Never," she replied. Always the amateur anthropologist, I pressed her to find out why. But all she would say is that it "wasn't right." I took it to the extreme, and asked her what she would say if the nicest and most physically attractive white man in the world was to ask her for a date.

Her response?

"Nope."

No logic. No reasoning. Just "Nope." She also admitted that she believed interracial dating in general should be discouraged.

I don't have much patience for those who act as though they are incapable of escaping the racist beliefs that were imposed on them early in life by their family and culture. Anyone can break free of such nonsense. It is possible to grow up in a race-obsessed culture—filled with all the standard prejudice, fear, and hate—and then just walk away from it all. I know, because I did it. It doesn't take great wisdom or courage. More than anything, it takes nothing more than the honesty to recognize the difference between what is real and what is make-believe.

The pace of change in attitudes in the United States is remarkable. According to Gallup polls, only 4 percent of Americans said they approved of marriages between whites and blacks in 1958. A staggering *94 percent* of Americans disapproved of black-white interracial marriage in that year. The remaining 2 percent had no opinion. By 2007, however, only 17 percent of Americans disapproved of marriages between whites and blacks, and 77 percent approved. Overall, 85 percent of Americans between the ages of eighteen and forty-nine approve of marriages between blacks and whites. Among those aged fifty and older, 67 percent approve. These results are similar among white Americans, as 86 percent of whites aged eighteen to forty-nine approve, compared with 64 percent of whites who are fifty and older.[40]

While it's encouraging that increasing numbers of people are dismissing traditional racial borders as less important or unimportant, America and the rest of the world still have a long way to go. As interracial marriage researcher Root points out, these types of relationships are not going to save the world. Nobody is suggesting that everyone needs to run out and date someone from another racial category. What sensible people are hoping for is the day when no one cares about this stuff, when all people look at humanity without blinders on, fall in love with anyone they want to, and then try their best to live happily ever after. The day we are hoping for is the day when interracial marriages are no longer worthy of celebration, research, or notice. We'll know we're there when interracial love finally is seen for what it always has been, simply love.

NOTES

1. Church of True Israel, "Let's Eat Your Son Today, Mine Tomorrow," http:// www.churchoftrueisrael.com/verboten/vb-11.html (accessed July 18, 2009).

2. Doug Linder, "Loving Et Ux v. Virginia," *Exploring Constitutional Law*, http://www.law.umkc.edu/faculty/projects/ftrials/conlaw/loving.html (accessed July 18, 2009).

3. "AAPA Statement on Biological Aspects of Race," *American Journal of Physical Anthropology* 101 (1996): 569–70.

4. Maria P. P. Root, *Love's Revolution: Interracial Marriage* (Philadelphia: Temple University Press, 2001), p. 9.

5. Ibid.

6. Associated Press, "After 40 Years, Interracial Marriage Flourishing," MSNBC, April 15, 2007, www.msnbc.msn.com/id/18090277 (accessed September 11, 2009).

7. Pew Research Center for the People & the Press, "How Young People View Their Lives, Futures and Politics: A Portrait of 'Generation Next,'" January 9, 2007, http://people-press .org/reports/pdf/300.pdf (accessed September 11, 2009).

8. *American Experience*, "Timeline: The Murder of Emmett Till," PBS, www.pbs.org/wgbh/amex/till/timeline/timeline2.html (accessed September 11, 2009).

9. Ibid.

10. Archives at Tuskegee Institute, "Lynchings: By State and Race, 1882–1968," University of Missouri–Kansas City School of Law, www.law .umkc.edu/faculty/projects/ftrials/shipp/lynchingsstate.html (accessed September 11, 2009).

11. Gunnar Myrdal, *An American Dilemma: The Negro Problem and Modern Democracy* (New York: Harper, 1944), p. 606.

12. Ibid.

13. Ibid.

14. Ibid., pp. 606–607.

15. Douglas Martin, "Mildred Loving, Who Battled Ban on Mixed-Race Marriage, Dies at 68," *New York Times*, May 6, 2008, www.nytimes.com/2008/05/06/us/06loving.html (accessed September 11, 2009).

16. Ibid.

17. US Supreme Court, *Loving v. Virginia*, 388 US 1 (1967), opinion delivered by Chief Justice Warren, http://caselaw.lp.findlaw.com/ (accessed September 11, 2009).

18. Martin, "Mildred Loving."

19. Annette Gordon Reed, ed., *Race on Trial* (Oxford: Oxford University Press, 2002), p. 187.

20. Martin, "Mildred Loving."

21. National Public Radio, "Loving Decision: 40 Years of Legal Interracial Unions," June 11, 2007, www.npr.org/templates/story/story.php?storyId=1088 9047 (accessed September 11, 2009).

22. US Supreme Court, *Loving v. Virginia*.

23. Theodore G. Bilbo, *Take Your Choice: Separation or Mongrelization* (Poplarville, MS: Dream House, 1947).

24. National Public Radio, "Loving Decision."

25. US Supreme Court, *Loving v. Virginia.*

26. Melia Patria, "Groundbreaking Interracial Marriage," ABC News, June 14, 2007, http://media.abcnews.com/US/Story?id=3277875&page=1 (accessed September 11, 2009).

27. Martin, "Mildred Loving."

28. Ken Tanabe, interview by the author, April 6, 2009.

29. Ibid.

30. Ibid.

31. American Academy of Child and Adolescent Psychiatry, "Multiracial Children," *Facts for Families* 71, updated October 1999, http://aacap.org/page .ww?name=Multiracial+Children&section=Facts+for+Families (accessed September 11, 2009).

32. Mike Stuckey, "Multiracial Americans Surge in Number, Voice," MSNBC, May 28, 2008, www.msnbc.msn.com/id/24542138 (accessed September 11, 2009).

33. US Census Bureau, "An Older and More Diverse Nation by Midcentury," August 14, 2008, www.census.gov/Press-Release/www/releases/archives/ population/012496.html (accessed September 11, 2009).

34. Karyn Langhorne Folan, "What Mildred Knew," *Washington Post,* June 12, 2008, p. A23.

35. Root, *Love's Revolution.*

36. Ibid.

37. Ibid.

38. Ibid.

39. Ibid.

40. Joseph Carroll, "Most Americans Approve of Interracial Marriages," *Gallup,* August 16, 2007, www.gallup.com/poll/28417/Most-Americans-Approve -Interracial-Marriages.aspx.

Chapter 7

RACE ON THE MIND

The problem with debates like the one over race and IQ is that psychologists on both sides of the controversy make a totally unwarranted assumption: that there is a biological entity called "race."
—Jefferson M. Fish and Ken Schles[1]

For the race difference in IQ, we can be confident that genes play no role at all. Most of the evidence offered for a genetic component to the race difference is indirect and readily refuted.
—Richard E. Nisbett[2]

Imagine two teenagers who are the same age. Both of them are generally healthy and have similar physiques. One of them, however, has spent the last ten years eating junk food and avoiding all forms of physical activity, while the other has eaten a nutritious diet and exercised daily, including lots of running.

Would it make any sense to have these two people compete in one race and then use the result of that single competition to declare one of them the *genetically* superior runner? Of course not. It could easily be the chronic couch potato who holds the superior genes for athletics. The fact that those genetic gifts have been buried by a lifestyle that is less conducive to running fast does not mean they were never there.

If the reigning Mr. Olympia is standing next to a skinny guy in a wheelchair, we cannot know with absolute certainty based on simple observation that the man in the wheelchair is the one with the inferior genes for bodybuilding. Perhaps he was on course to be better than Mr. Olympia but was injured in a car crash ten years ago. This is the problem with intelligence tests. Measuring performance on one test at one moment in time does not necessarily measure broad current ability, potential ability, or what the test taker might have been capable of given different life experiences leading up to the day of the test. Nevertheless, for a long time, many people have been trying to do just that. Many people believe, for example, that we can give a test to some black people and some white people, average the scores, and come up with racial winners and losers. Proponents of this idea claim that such tests not only show who is "smarter" but also which entire race is smarter. While some scholars are screaming that this is dangerous nonsense, others continue to promote this line of thinking. Meanwhile, millions of people—including some people in positions of great power—use these data as justification for all the racial disparities they see in their society and around the world.

"So black people score below white people on IQ tests," reasons some politician or bureaucrat. "Well, that helps explain black crime in America and all those starving children in Africa. They just aren't very smart, and nothing can change that."

This is a terrible misreading of reality, which carries with it terrible consequences. Believing in a mostly unchangeable innate intelligence that is largely determined by "race" means not believing in the biological reality of our species. Worse, for some people it can mean not having to care as much about children who are neglected and denied decent education, security, and nutrition. Why care? Their genes have doomed them to an existence at the bottom of society anyway. What's the point of trying to help them? Fighting against racism and trying to eliminate racial inequalities won't matter in the end.

In a perfect world, one where all children were born into a safe place with support and countless opportunities before them, it might make sense to draw conclusions about IQ tests (assuming they are not culturally biased, of course). But our world is far from perfect. And I don't mean some kids get to Harvard and some kids have to settle for junior college. Millions of children *die* each year, only because they were born into poverty. Millions more who survive into adulthood suffer the ravages of poverty, both physical and mental. For example, growing up poor and suffering the daily grind of stress that it brings has been found to impair working memory later in life.[3] Gee, you think that might make it a little tougher for some poor kids to ace an IQ test and score high enough on their SATs to get into Yale? How can anyone be so cold and thoughtless as to believe that it is fair to test people who grow up in deprived and dangerous conditions—who happen to be disproportionately nonwhite, by the way—and then draw conclusions about their *innate* intelligence, as if the extraordinary challenges of their life experiences mean nothing? Furthermore, how can we justify trying to compare the innate (biological) intelligence of races when so many scientists today tell us that biological races do not even exist? Why don't we try to determine the innate intelligence of people born on a Tuesday versus people born on a Thursday and then make a big deal about that too? There are bound to be tantalizing differences between them if we look hard enough and long enough.

"The fallacy is not that we have positive knowledge that different people (or groups) have identical abilities," states biological anthropologist Jonathan Marks. "Rather it is that we have no way to know what their abilities are, except in retrospect, after those abilities have been cultivated to some extent. This results in a basic asymmetry in the relation between performance and abilities."[4] It is important for the layperson to understand this. Scientists and psychologists are not measuring "brains." (They tried that a long time ago and it didn't work.) They are giving people a test. That's it. Imagine, not just a grade or your educational future riding on the result of a single test, but the judgment of your entire race as well. Yes, the perceived intelligence of hundreds of millions of people who lived before you, who are alive now, and who have not yet been born will be helped or harmed by your test score. What if you have a bad day?

"Even assuming that one buys into the existence of biologically distinct 'races,' the roughness of IQ tests ought to lead, at a minimum, to agnosticism on the subject," writes Stephen Murdoch, author of *IQ: A Smart History of a Failed Idea*, an excellent survey of the weird and often tragic story of intelligence testing.[5] "Until people live in equal conditions in terms of income and health, and there exist far more exact mental measurement tools than IQ tests, there should be a collective shrugging of shoulders. Even then, it's not clear why the subject would be particularly interesting or useful. [Some] psychologists continue to pursue this nauseating inquiry, however, often becoming defensive when asked why it's important to study, but not coming up with satisfying answers to the question.... The obsession is strange and, at times, harmful."[6]

Of all the issues revolving around the concept of race, intelligence is the most important. The idea that humankind can be sensibly divided up into biological categories that can then be ranked by intelligence has been and continues to be at the root of a great deal of trouble. The claim that members of some races are incapable of thinking as well as members of other races, because of inferior genes, is not just unproven; it is unconscionable. This idea has been used to justify prejudice, segregation, slavery, sterilization, and even extermination. It has been the traditional explanation or justification presented for the mass theft and genocide perpetrated by white people against indigenous people in the New World. And it remains a popular explanation offered for the stark inequalities we see in the world today. The claim that blacks, Hispanics, or Native Americans, for example, are doomed by DNA to inherit lower-class status generation after generation must be an attractive proposition for those who might otherwise feel uncomfortable having to explain an unfair society that they benefit from.

The race-intelligence link is cited out loud or assumed silently as the reason for the differences in the way racial groups fare educationally, socially, and economically. According to conventional wisdom, white Europeans won more often than not throughout history because they were the smarter race, the people with greater innate intelligence. President Barack Obama aside, white people still dominate most of the world's most powerful governments, militaries, economies, and research centers. White people still win most of the Nobel prizes. On average, white people in the United States still live in the best neighborhoods, have the best jobs, and send their children to the best schools. Some say that slavery ended long ago and that a little discrimination here and there can't explain the racial stratification we see every day. The only sensible explanation is that white people must have higher natural intelligence and thus an advantage over black people, Hispanics, and Native Americans. (Interestingly, Asian

American immigrants faced the same assumption in the nineteenth century. After a few generations of remarkable academic success by Asian Americans, however, people with racist beliefs have had to modify their perceptions.) White people are born smarter, so they end up doing better. Isn't this one racist claim that is supported not just by casual observation but also hard scientific data? After all, blacks, for example, come up short on IQ tests. There is no disputing the numbers, right?

Wrong.

There are plenty of reasons to doubt if not outright reject the notion that race and inherited intelligence are tied together in any way that would allow us to rank races. As this chapter will show, there are significant problems with the racist view of intelligence and with intelligence testing itself. While it is clear that many people, probably most people, believe in a meaningful connection between race and intelligence, this chapter will show that this belief is unfounded.

Popular beliefs and opinions aside, what does science really say about race and intelligence? Countless very smart people have looked into this question over many years. What the slaveholders and Ku Klux Klan members assumed to be true, scientists sought to prove. After centuries of effort, however, there is virtually nothing substantial to show for it. Shouldn't we have some definitive answers by now? If there is anything to the connection between race and intelligence, where is the proof? Can we identify, measure, and rank inherited racial intelligence or not? Can we stack human groups by innate mental ability, based solely on hard data and honest discovery, free of corrupting influences such as bias and racism? After measuring craniums, endlessly poking and prodding people, and testing millions of children, what can we confidently claim to know today? What can we be sure about, after so many scientists have devoted their careers to proving once and for all that white people are smart and black people are not?

Sorry, still no answer. Some would say the game is over and the race-intelligence claim has lost, but there are still many genuine believers as well as some researchers who are working hard trying to prove it, so the issue remains unresolved. However, those who make the claim of a meaningful connection between inherited intelligence and race are the ones who must prove their case. As we shall see, they have failed to do so thus far.

An honest, reasoned assessment of the race-intelligence issue inevitably leans toward the racist argument as being at best unproven and at worst flat-out dishonest. This is a critically important point that needs to be made clear to the general public: *Science has not proven that race and inherited intelligence are linked.* It is important for people to hear this message, because so many wrongly assume that science has proven that racial intelligence can be ranked. I recall once chatting with a bright woman who later earned a PhD. We were talking about the places we had visited on our respective travels, and poverty came up. Much to my surprise, she said that, while she was upset about injustice and inequality in the world, she felt that this problem was inevitable. After all, it "has been proven that black people are mentally inferior." When I questioned her about this scientific proof, she cited "IQ tests."

This is typical of the sort of thoughts that are rattling around in the heads of good and otherwise sensible people all over the world. For one reason or another, they have latched on to the idea that credible scientific work has confirmed the fact that races vary in inherited intelligence. They seem to believe that everyone on Earth has been given an IQ test, or at least enough people to enable sound conclusions to be drawn. They think that intelligence tests are not biased in any way and that they actually do measure innate general intelligence. They also assume that biological races exist and something like racial intelligence is real. As we shall see, all of these assumptions are faulty, unproven, and very dangerous.

Notice, by the way, that I often insert "innate" or "inherited" in front of the word "intelligence" when writing about the idea of racial intelligence. This is important, because this discussion is not simply about people of different races differing in mental ability or academic performance in school. Of course there are differences. Compare any two groups of human beings who have been treated differently by society and you are likely to find differences. Also, individuals can and do inherit unique intellectual potential. We are not all born with exactly the same brains or genetic potentiality for complex thinking. However, if a group of people—a race, an economic class, an oppressed religious minority, and so on—is denied the education, security, nutrition, healthcare, and positive social environment that another group enjoys, then it is unlikely that the two groups will have identical average test scores. Denying that a difference in test scores between two racial groups is the result of inferior genetics is not the same as denying that the difference may exist.

Races, whether one accepts them as biologically real or not, can have profound effects on virtually every aspect of a person's life. For example, if on March 14, 1879, a black boy was born in Mississippi with a brain that was every bit as gifted and packed with potential as the one Albert Einstein was born with that same day in Germany, I am pretty sure they would not have tested the same on an IQ test ten, twenty, and thirty years later. The black boy might do relatively well in his neck of the woods, but it is difficult to imagine that he would end up even with Einstein in adulthood, despite their original genetic equality.

It is obvious that people fare better or worse because of the differences in the environments, expectations, limitations, and opportunities that they are born into. No one should dispute that. But some people seem certain that these factors are minor compared to inherited *group* ability. This is an amazing claim when we stop and think about it. Significant numbers of black people and Native Americans don't struggle to get ahead, for example, because of the way they have been treated for the past few centuries and are still treated today. They struggle because they can't do any better; it's just not in them. Yes, people really do believe this. One can see the attraction. It's the perfect escape from historical guilt and current responsibility. Pizarro and Cortés didn't so much slaughter Native Americans as facilitate their descent into their rightful place in the world's racial ranking. Slavery, segregation, and discrimination in the United States didn't so much hold back black people as help to familiarize them with their role in society as predetermined by nature long before the United States even existed. Every time someone helps spread the idea that science has "proven" the mental inferiority of some races, this is the sort of thinking that results. The potential harm that comes from promoting such ideas is clear. Therefore, no responsible scientist or layperson should ever suggest that a scientifically validated link between inherited intelligence and race exists. It has simply not been proven by any reasonable standard, so it should not be tossed around in public for racist minds to feed on.

THOSE WHO HOLD THE POWER GET TO WRITE THE TEST

The issue of intelligence tests is key to the question of race and inherited intelligence because this is where a great deal of research has been focused and it is something that the public easily relates to. An IQ score, that single number, makes it all so easy to understand. It also makes it seem scientific and official. If Asians score higher on intelligence tests than whites, and whites score higher than blacks, then "That's enough for me," says the average person on the street, "Some races are just smarter than others."

Although much is made of racial groups performing differently from other racial groups, it is far from proven that these differences are based on genetic differences. It is sad that the IQ score has become so damning for some and so divisive for all. "How strange that we would let a single and false number divide us, when evolution has united all people in the recency of our common ancestry—thus undergirding with a shared humanity that infinite variety which custom can never stale. *E pluribus unum*," wrote the late evolutionary biologist Stephen Jay Gould.[7] I attended one of his lectures in the 1980s and never forgot his warning always to be aware of who is doing the measuring, and the testing, and the concluding when it comes to matters of science. Gould loved science, of course, but he was well aware that as long as humans are the ones doing it, science will be imperfect and subject to human biases—often from the subconscious. This, I suspect, is how we end up with so many highly educated white people giving tests requiring a knowledge of Latin, algebra, and essay composition to poor black and Hispanic children who attend failed inner-city schools, and then assuming that inferior genetics are the reason if these children don't perform as well as wealthy white children in Vermont.

For many years now, white Americans have outscored black Americans by about fifteen points on IQ tests. (whites: 100; blacks: 85) Racists have gotten plenty of mileage out of this. But is it because of genes or because of social conditions? Do blacks really lag behind on IQ tests because of evolutionary events tens of thousands of years ago, which determined their innate intelligence as a people? Some say the gap is simply too large to be explainable as the result of racism and discrimination. It must be genetic, they say.

One of the biggest problems for those who believe in the race-IQ link is that IQs for *everyone* have risen rapidly in recent decades—too rapidly. There has been a gain of some fifteen points in the last fifty years or so that cannot be explained as genetic, because it has happened far too fast. In the United States, the average IQ for the entire population rose eighteen points from 1947 to 2002. It must be environmental, the result of better education, cultural changes, health improvements, or more opportunities. "Genes could not have changed enough over such a brief period to account for the shift," writes University of Michigan psychologist Richard E. Nisbett. "It must have been the result of powerful social factors. And if such factors could produce changes over time for the population as a whole, they could also produce big differences between subpopulations [such as racial groups] at any given time."[8]

The rapid, planet-wide rise in IQ scores is called the Flynn Effect, named after James Flynn, the prominent psychologist who discovered it. Flynn is a friendly fellow who has dedicated much of his career to researching intelligence as it relates to race. In my exchanges with him, I got the feeling that he is a compassionate person who cares about people in general, but not to the point that he would be willing to compromise whatever reality he might uncover in his work.

Racial gaps in intelligence scores "are more likely to be environmental, but no one can make a final judgment until all of the evidence is in," said Dr. Flynn. "I believe this because of the IQ gains blacks have made on whites in recent years and because the life histories of blacks in America today hardly match the quality of the average white. The Flynn Effect only shows that environment is a much more potent influence on IQ than we had previously believed. It does not save us the trouble of careful study of the life histories of black and white and isolating exactly what in the black environment is harmful."[9]

Flynn warns against seizing his work as final proof that there is no race-gene-intelligence link. He says nothing can be taken for granted. "The fact that blacks have cut the IQ gap by a third in the last generation shows that those who thought the whole gap was genetic were wrong. I think I have made a strong case that the rest is environmental, but people will have to evaluate my case for themselves."[10]

Flynn believes racial groups that have struggled need to do more for themselves. Social policies and wise investments matter, of course, but what happens in the home is also vital. "Black parents are already nurturing, in the sense that they love their children, but they have to go beyond that to give them a home with greater cognitive challenge," says Flynn. "Children are more likely to read serious books if they actually see their parents reading. School alone cannot educate if your teenage peer group rejects what school has to offer."[11]

In his thought-provoking book, *Where Have All the Liberals Gone? Race, Class, and Ideals in America*, Dr. Flynn is blunt about this: "I believe that they [black Americans] tend to live in a distinctive black subculture that offers a less rich spectrum of cognitive environments at every stage. And the quality of environments at every stage influences the quality of environments available at the next stage."[12] Dr. Flynn guesses that black Americans today are about where his social group, Irish Americans, were in the 1920s. "But we need not be that pessimistic!" he says. "They should learn from our mistakes."[13]

Flynn points to a fascinating study of children who were fathered by American soldiers in Germany during the US occupation after World War II. Something about growing up in Germany and attending school there must have been very different from growing up in America and attending American schools, because the average IQ scores of children with black American fathers were virtually identical to the scores of children with white fathers. The psychologist who did the study, Klaus Eyferth, found that the children of black fathers outperformed the white children on many sections of the intelligence test, including picture arrangement, information, and comprehension. Overall, the children with black fathers had an average score of 96.5 and the other children averaged 97.0. This difference is statistically insignificant.[14] "The American black environment must be damaging in some way that the German black environment was not," Flynn concludes.[15]

Another study, similar to Eyferth's in Germany, found that black children raised in an English orphanage scored higher on intelligence tests than white children who were raised there.[16] It is clear that there is sufficient reason to stop short of confidently pointing to genetics as the cause of differences in IQ scores between racial groups.

The Flynn Effect is widely believed to be a result of better conditions and opportunities. "I think the obvious explanation for the Flynn Effect is education," concludes St. John's University emeritus professor of psychology Jefferson M. Fish. So, if environmental factors such as improved schooling, nutrition, and/or better test-taking skills account for the recent rise in global IQs, then why can't they account for the gaps between races, especially when members of some races live in very different environments and have very different experiences with racism and discrimination? Also, why did black Americans score better on intelligence tests when they moved from rural homes in the South to urban homes in the North?[17]

Genes related to innate cognitive ability do not change magically just because a person crosses the Mason-Dixon Line. The only sensible assumption is that something changed in the environment; maybe it was better schools and less racism, for example. Black children who are adopted into wealthier and higher status homes also tend to improve their scores.[18] And, finally, why is it that a black woman in America is twice as likely to have an IQ above 120 than a black man? They are both black. Same race. There is no such difference between white male and white female IQs. The most likely explanation is that something in American culture and/or within the black subculture damages black male intellectual development and educational achievement more than it damages intellectual development and achievement among black women.[19]

We know that scores on intelligence tests and achievement tests can improve when environmental conditions improve. Why, then, is it so difficult for some people to accept the likelihood, or possibility, that the horrendous environmental conditions black Americans have suffered under for centuries, as well as the below-average conditions so many still live in today, are likely to be responsible for that black-white gap in average IQ scores? The life experience of many nonwhite children does not position them for a good showing on a test created by people who dwell in a world that is very different from theirs. For some children and teenagers, being given a standardized intelligence test probably feels like making contact with an alien species.

Before we go too deeply into intelligence testing and its impact on race, it is important for us to remember that, according to most anthropologists, biological races do not exist. Races are cultural creations. It is important to keep this in mind, because when most psychologists talk and write about races, they tend to do it in a way that assumes races are logical and scientific categories of people. This shouldn't be surprising, because, after all, psychologists are people too, and that means they are susceptible to mistaken beliefs just like anyone else. It is important to understand that one can make a very good argument for discounting all IQ test data and conclusions as they relate to race and inherited intelligence simply because races are not real. For example, if a psychologist gives an intelligence test to a "black American" and then feeds that score into a database to be averaged with thousands of other "black American" IQ test scores for the purpose of drawing conclusions about genetics or inherited intelligence, the data are already suspect and flawed, because the original identification of the test taker is suspect and flawed. It's just not good enough for a psychologist or teacher to look at a test taker and assign a racial label. Nor is it good enough for the test takers to label themselves. Why? Because there is no genetic or scientific sense to racial labels, especially in a society like America. A "black American" for example, can have an Asian or white parent but still "look" black and identify herself or himself as "black." This is culture at play, certainly not biology. This is why comparing the IQ averages of various races in the United States to prove a point about inherited or genetic group intelligence is problematic if not fatally flawed. "The question 'How do you know what race your participants are?' has not been understood by psychologists to be unanswerable," writes psychologist Fish, who is the editor of *Race and Intelligence: Separating Science from Myth*, an important book about the race and intelligence issue.[20] "Meanwhile, the classification of people into biological races has long been known by anthropologists to be scientifically inaccurate, but reflective instead of American folk beliefs, which differ from folk beliefs in other cultures. Because psychologists in the United States are culturally American they take these scientifically inaccurate beliefs for granted. When they assume that human races exist, and categorize their experimental participants by physical appearance or self-designation, they unwittingly create much mischief."[21]

The end result is a bunch of researchers producing influential work that is believed to be about biological groups but is really about cultural groups. The error some American psychologists have made and continue to make is that they take the "commonsense," "plain-to-see" stance toward race. "Of course it makes sense to compare differences between racial groups, because racial groups are obvious," a race-believing researcher might say. "Anyone can pick out people on any street and identify their race." But this is not so, according to Fish. He points out that easy identification of the races of strangers on a random American street only works if it is an American who is doing the identifying. The American mode of racial identification doesn't work in other cultures, because the rules may change dramatically. In Brazil and Haiti, for example, racial classification is radically different from America's "obvious" system of grouping people. A typical American would have no chance of going to São Paulo or Port-au-Prince and accurately identifying strangers by race in the way that the local people would do it. Likewise, a typical Haitian or Brazilian would likely struggle, at first at least, to identify races on an American street in the American way.[22]

"In almost all studies the so-called racial background of individual respondents and respondent populations has been derived in ways that show no resemblance to means used by genetic specialists," writes Yale psychologist Edmund W. Gordon.[23] He continues:

> In those few cases where any information is given about criteria of assortment, one usually finds that skin color has been the sole or dominant criterion, and that as measured by the eye. In other words, the actual genetic background of the subjects is uncontrolled. The most useful studies linking race and certain specified socially valued traits make no pretense of dealing with biogenetic race: rather, they openly work with categories of "social race."...If race is to be treated as a sociocultural construct, it is important to get the individual's views on his own identification and the identification he applies to others. However, if race is to be treated as a biological construct, the lay individual's views of his own racial identity or [those] of anyone else are unqualified and immaterial.[24]

A FAILURE TO COMMUNICATE

During a long conversation about race, psychologist Jefferson Fish stressed to me that there is a serious problem in his field as well as within academia in general. "Once you get outside of biology and anthropology, you see that the word [about the nonexistence of biological races] just hasn't gotten around. One of my main messages is that psychologists need to learn some anthropology," he says.[25] "The main problem is that people are not learning what is going on outside of their field. Race can be frustrating because it's such a deeply ingrained cultural belief. There is this desire to use race because it allows researchers to generalize. It's convenient. But it just doesn't work in too many cases. For example, if you wanted to compare 'whites and blacks' on an IQ test in Brazil with 'whites and blacks' in the United States, you couldn't do it, because people who are called 'black' in the United States wouldn't be called 'black' in Brazil and people who are called 'white' in Brazil wouldn't be called 'white' in the United States."[26]

Dr. Fish says that anthropologists figured out that biological races are not real decades ago, and toward the end of the twentieth century they began scratching their heads, wondering why nobody else knows it. "Psychologists just don't know this stuff," he explained. "If you look in psychology journals, sociology journals, and other scientific journals, you will see that they still use terms like 'Caucasian.' It's just ridiculous. The first problem is that psychologists never heard that biological races are not real. The second problem is that often when it comes up, it is turned into a political debate rather than a scientific inquiry. And the third problem is that psychologists like to generalize to all humans, and race is a handy concept for that."

The manner in which Fish, a New York psychologist, came to be so enlightened on the subject of race was mostly accidental. "I just happened to have married an African American woman who was an anthropologist," said Dr. Fish. She was up on the biological nonexistence and social reality of races. He was not. The first few years of marriage were an enlightening experience for him. "Then we went to live in Brazil for a while. I did some reading on race, and I saw how the concept of race is very different there from the way it is in the United States. I was in an almost constant state of amazement about all of this."

Unfortunately, not everyone can marry an anthropologist. Dr. Fish is hopeful that education about what races are and are not will improve among researchers and especially among younger children. As an educator, Dr. Fish has great confidence in the potential for improving race understanding through education as well as through better communication between fields of study. However, he does not believe that this will necessarily stop people from separating themselves and hating one another: "Unfortunately, there are so many things we use to divide ourselves up… that it seems like it will never end. But that is no reason for us not to try and do a better job with the concept of race."[27]

SHOULD WE STUDY INTELLIGENCE AND RACE?

I suspect that there are some individuals who sincerely care about all people and disapprove of racism but shy away from talking about the issue of race and intelligence because they fear there may be something to it. In the minds of some, it's a can of worms best left unopened, a minefield not worth traversing. I disagree. I believe we should pursue every possible discovery about ourselves, even if it may potentially lead to uncomfortable truths. The more we know, the better the chance for us to make sensible decisions and move forward in the best direction. Besides, how much trouble could awareness and truth get us into? On race, ignorance and lies have not served us very well to date. For the last several thousand years, we have relied on folk tales, fear, religious beliefs, and pseudoscience to inform our views and behaviors regarding our biological diversity. The many hundreds of millions of dead people and the many hundreds of millions more who endured degraded and limited lives make the point that ignorance offers no guarantee of peace and harmony. Yes, scientists have promoted terrible ideas about race that have contributed to the destruction of human lives. But the saving grace of science is that it is self-correcting. Scientists, after all, were the ones who eventually declared that phrenology was pseudoscience. And scientists today are leading the charge against belief in a connection between race and innate intelligence.

Another reason why I support the pursuit of any and all knowledge about intelligence is my confidence that it will not, as long as we remain honest with regard to the evidence, lead us to a Nazi renaissance and a KKK utopia. Biological races are not real, so it seems unavoidable that sooner or later we will find that there are only insignificant differences between the genetic intellectual potential of large groups of human minds, no matter what color bodies they happen to be in. Yes, our environments, histories, and opportunities as well as our individual genetic gifts and curses are different. But those things are not the same as unique intellectual potential determined by imagined biological races. We are such a jumbled, closely related, young and blended species that we can not justly or accurately make sweeping judgments about this group or that group without erring every time. In the end, I am confident that we will learn only that there is nothing to be gained by trying to link mental potential to groups of humans lumped together in something so illusory and fleeting as races.

We have been obsessed with unlocking the mysteries of human thought for a long time. Much effort has been made, not just to understand the minds of people, however, but to measure and draw restrictive boundaries around people as well. In previous centuries, craniometry and phrenology were the preferred methods for ranking human brainpower. Scientists rubbed heads and measured skulls with great enthusiasm and dedication in order to determine who had the biggest brains (meaning who should rule society and who should do janitorial work for the big-brained race). Of course, the skulls of white people always seemed to come out on top, a result that fitted nicely with the worldview of the white scientists who were doing the measuring.

Although it is still not completely gone from the face of the Earth, phrenology has been thoroughly discredited and rejected by mainstream science. Although physical anthropologists and forensic anthropologists still measure skulls, virtually no scientists today accept craniometry as a valid means of ranking racial intelligence in modern people. But while rubbing and measuring heads has faded, intelligence testing has exploded. Who needs to get their hands dirty massaging the oily heads of strangers? Who needs calipers? With little more than paper, a pencil, and a stopwatch, one can measure a person's intelligence. Collect enough scores and you can measure the intelligence of an entire race. Many individuals and institutions have been quick to believe this.

A hundred years ago, some psychologists saw intelligence testing as the Holy Grail. Finally, they had found a yardstick by which men, women, and children could be measured and sorted into categories. A genius here, a coal miner there: it was an irresistible idea. How useful it would be, they thought, to be able to administer a standardized test—even to very young children—and, within no more than an hour or so, not just determine current mental ability and arrive at a prediction of future success or failure, but also determine the subject's inherited or innate intelligence. Furthermore, with a score, this numerical value of a person's intelligence, one could add up and average scores to find out which gender, groups, nationalities, and races were smarter. A nation's immigration policies could be built around such tests. Business owners could weed out dimwits and hire geniuses, or simply rely on racial averages to inform their decisions. Schools could be fortified with these tests, keeping out students with inferior minds or inferior racial pedigrees. Test results like these could even be used as evidence to argue against programs and policies designed to remove the social injustices and inequalities suffered by some racial groups. Why pay for that preschool when "those" people have below average intelligence to begin with?

The roots of intelligence testing reach back to Alfred Binet and Francis Galton in the early twentieth century. The Frenchman Binet and his colleague Theodore Simon are credited with producing the first intelligence test. Their purpose, however, was far from that of ranking races and individuals by innate intelligence. The test was designed specifically to identify students who had learning disabilities or special challenges that required additional attention—nothing more. A few years later, however, Stanford psychologist Lewis Terman modified Binet's test, turning it into something that he claimed could measure anyone's intelligence. What is commonly known as the IQ test was born.

The crucial moment that put intelligence testing on the map and gave these tests lasting popularity and power occurred during World War I. When war broke out, American psychologists were a "bickering and unfocused lot, like small uncoordinated boys left on the sidelines of a pickup basketball game."[28] While other fields of science were helping with America's war effort, psychology had not yet figured out what its contribution would be. Seven men changed that in a smoky Philadelphia hotel room on April 21, 1917. They decided to give the army a test that could grade and sort human minds.

"Despite the meeting's obscurity, both then and now, it would be surprisingly important to the field of psychology and, in years, the entire world," explains Murdoch in *IQ: A Smart History of a Failed Idea*. "Of all the ways psychology could have contributed to the war effort, these men successfully focused the field's attention on a limited, narrow kind of intelligence testing that would catch on after the war.... We are still feeling the consequences today."[29]

A handful of American psychologists successfully sold the US Army on the idea of a test that could be given to soldiers in order to help decide what they were best able to do as well as eliminate those not suited for the military. Terman believed that while the German army was homogeneous and efficient because it was made up of one race, the US Army was an "assembled horde" made up of too many races. What America needed was an intelligence test to achieve organization and efficiency.[30]

Murdoch describes Terman's claims and attitudes:

> Despite the scholastic content of his questions, Terman didn't believe that his Stanford-Binet [test] tested students' educational and cultural background. He claimed that his test was able to isolate and measure innate intelligence—a fixed, inheritable trait. It was a supposition that had clear political and social implications, as revealed in his description of two low-scoring Portuguese boys who "represent the level of intelligence which is very, very common among Spanish-Indian and Mexican

families of the Southwest and also among negroes. Their dullness seems to be racial, or at least inherent in the family stocks from which they come. The fact that one meets this type with such extraordinary frequency among Indians, Mexicans, and negroes suggests quite forcibly that the whole question of racial differences in mental traits will have to be taken up anew and by experimental methods.[31]

The tests used by the US military in World War I were flawed, they were not very useful to the military, and they inspired incorrect conclusions—but they didn't go away.

"While the tests may not have helped the American war effort," concludes Murdoch, "there were four extremely important consequences. One, psychology established its reputation on the basis of the exams. Two, whereas before the war mass testing could have taken many different forms, including tests of many personality characteristics, psychology rallied mainly around measuring intelligence in a single, rigid way. Three, because of the use of tests during the war, American schools would flock to intelligence testing, and a lucrative testing industry would be born. And four, analyzing Alpha and Beta exam results became an intellectual cottage industry in the 1920s that bolstered racist, nativist, xenophobic tendencies."[32]

The United States was not alone in embracing intelligence testing, of course. British tests, such as those that were part of the eleven plus, were used not just to test knowledge but also with the belief that they measured innate intelligence.[33] Murdoch declares:

Thus IQ test technology which grew out of practical and historical necessity (not from knowledge of or agreement about what intelligence actually is), hasn't changed much in the hundred years since Alfred Binet published his first test, in 1905. Change has occurred, but often in formalistic ways, such as the use of multiple-choice questions for group tests and cultural adaptation of the questions to America and elsewhere. The decades-long structure of verbal and performance questions in the Wechsler tests and IQ exams does not come from intelligence or cognitive theory but from historical legacy, statistical relationships between how people score on tests and subtests, and some power to predict future behavior. Psychology's resistance to change is what keeps this structure alive.[34]

It is important to keep in mind that Binet (the creator of the test that Terman transformed into the Stanford-Binet test) never made the claim that his test or any other test could determine a person's inherited intelligence.[35] Nevertheless, it was Terman's test and Terman's view of what the results stood for that took hold in America and around the world.

Below are some of the questions included on the test given to US soldiers during World War I for the purpose of measuring their "innate intelligence." Remember, sweeping judgments about racial intelligence were also made based on the answers to these questions:

Informational Questions[36]

Pinochle is played with
a. rackets
b. cards
c. pins
d. dice

The Wyandotte is a kind of
a. horse
b. fowl
c. cattle
d. granite

Salsify is a kind of
a. snake
b. fish
c. lizard
d. vegetable

Rosa Bonheur is famous as a
a. poet
b. painter
c. composer
d. sculptor

Velvet Joe appears in advertisements of
a. tooth powder
b. dry goods
c. tobacco
d. soap

The bassoon is used in
a. music
b. stenography
c. book-binding
d. lithography

The scimitar is a kind of
a. musket
b. cannon
c. pistol
d. sword

The author of "The Raven" is
a. Stevenson
b. Kipling
c. Hawthorne
d. Poe

Spare is a term used in

a. bowling
b. football
c. tennis
d. hockey

Mauve is the name of a

a. drink
b. color
c. fabric
d. food

The Overland car is made in

a. Buffalo
b. Detroit
c. Flint
d. Toledo

Analogy

tiger–carnivorous :: horse–

a. cow
b. pony
c. buggy
d. herbivorous

Practical Judgment

Why are pencils more com-monly carried than fountain pens? Because

a. they are brightly colored
b. they are cheaper
c. they are not so heavy

Why is leather used for shoes? Because

a. it is produced in all coun-tries
b. it wears well
c. it is an animal product

Looking back, it is surprising that people actually thought the test measured inherited intelligence, when it so clearly measured general knowledge and test-taking skills, things heavily influenced by schooling and environment. But, then again, isn't that what many people still believe about intelligence tests today?

Could a test made up of questions like those listed above measure knowledge? Sure. Could it offer some reasonable prediction of success or failure in the military, in school, or in some kinds of jobs? Maybe. Could it determine a person's genetic intelligence or be used to determine the average innate intelligence of a racial group made up of millions of people? No chance. Whether or not one knows that a bassoon is a musical instrument or that Velvet Joe has something to do with tobacco advertisements is determined by cultural environment, not by some deep racial ancestry written into the DNA. In the early twentieth century, a white man from Manhattan who took this test would have had an advantage over a black sharecropper from South Carolina. To assume that a difference in their scores would be the result of racial genetics rather than environment and education is ludicrous. Yet millions of people—millions of black people, Hispanics, immigrants, women, and others—were judged and condemned by such questions.

Intelligence tests have changed since World War I, of course. But not nearly enough, say critics. The critics charge that they are still tests of knowledge that are biased in favor of the perspective of those who make the tests. The closer a test taker is to the culture of the test makers, the greater the advantage. You would think it would be obvious that a test like this heavily favors not only test takers who have been exposed to items that are on the test, but also test takers who have grown up in positive learning environments and attended good schools with good teachers. And don't forget the disturbing disadvantages some test takers have faced during their mental development, such as malnutrition and stress resulting from discrimination and fear of lynchings. A disproportionately large number of nonwhite Americans have grown up in educationally impoverished environments and attended poor schools with inferior teachers. It is outrageous to conclude that poor results on a test like the SAT (a direct descendant of the earliest IQ tests) are evidence of a mentally inferior race. Having a good vocabulary and being comfortable with analogy problems is unlikely to have anything to do with one's genetic ancestry stretching back to the time when humans diverged and populated different continents. On the other hand, it is likely to have everything to do with numerous immediate factors, such as the quality of schooling and the consistency of the at-home educational encouragement received or not received during childhood.

Many interesting facts arise in the course of researching IQ and race (some of which have been mentioned previously). For example, did you know that during the early twentieth century, blacks in the North scored higher on intelligence tests than blacks in the South? Another interesting tidbit is that, among black children who had migrated from the South to the North with their families, the longer they had been in northern schools, the higher they scored on IQ tests. Just as a result of moving and attending a northern school, black children's IQ scores went up significantly.[37] There is evidence that suggests IQ scores can be raised by 8–25 points and maintained with childhood intervention. One researcher found that, from the end of high school to the end of college, black students improved their test scores at a rate four times greater than that of other students.[38] I wonder how black American teenagers would have performed on the SAT test if it had existed back in the late 1800s and early 1900s. I'm guessing that there would be a tremendous difference between their scores then and now. But that difference would be purely environmental. In 1976, more than 30 percent of African Americans who took the SAT earned a score of less than 1000. By 1989, only about 12 percent did so. About 4 percent of African Americans scored 1300 and above in 1976. By 1989, that figure had more than tripled, rising to approximately 14 percent.[39] Interestingly, the average black American's IQ score today is equivalent to that of the average white American in the 1940s.[40] If the racists are right and blacks are mentally inferior, then what does that say about the revered "Greatest Generation," which won World War II? Furthermore, if intelligence tests measure inherited cognitive ability—something that is determined by genetics, evolution, biological race—then something very unusual is going on with the genes of America's black people. They must be evolving at an astonishingly rapid pace. (For those who may be unfamiliar with evolution, one hundred years or so is far too short a time period for evolution to account for massive changes in inherited racial intelligence. Evolution might move that fast with fruit flies and microbes but not large mammals like us.) Black people in America have shown the ability to improve their intelligence tests scores far too fast for us to attribute the improvement to anything other than environment. And if they can improve so rapidly, doesn't that mean their scores were low due to nongenetic (not biological race) factors in the first place?

BORN IN A FOREIGN LAND?

Psychologist Joseph Fagan does not believe that differences in black and white IQ scores are due to inherited or innate intelligence. He thinks language may be behind the differences. Because IQ tests have so much to do with world knowledge, Fagan wanted to see if that could be the key to the black-white gap. He tested three groups of people: whites, blacks, and white people who were not native English speakers. On an IQ test of vocabulary, the whites scored sixteen points more than the blacks and eighteen points more than the non-native English speakers. Fagan then gave the three groups a test that was written in "black English." The blacks answered 85 to 95 percent of the questions correctly, while neither group of white test takers could do any better than 40 percent correct. Finally, Fagan gave the groups a list of obscure words to study, words he predicted they were not likely to be familiar with. He then tested them on the words. While the scores varied among individuals, no group ranking based on average scores emerged.[41] On a neutral battlefield, the three groups were even. Just by tinkering with the language of the test, Fagan was able to completely flip the racial hierarchy and then make it vanish altogether.

LIVES IN THE BALANCE

As one who took an intelligence test as a young child and benefited from a good performance (admittance into a "gifted program") and took the SAT as an uninterested teenager and paid the price for a less-than-record-breaking performance (no Ivy League for me), I sympathize with all those countless millions of young people who may have choked, had a bad day, suffered from a headache, or simply weren't able to show off their true lifetime's worth of intellectual potential with a number two pencil in the allotted time. If one test is going to forever alter the course of a person's life, particularly in childhood or the teen years, doesn't someone have an obligation to show beyond a doubt that the test really does measure innate intelligence? If it doesn't, aren't we making a huge mistake in allowing people's lives to be judged and guided in this fashion?

"Why one person becomes a grocery store checkout clerk and another becomes a professor is fodder for astrologers, not scientists," writes biological anthropologist Jonathan Marks. "As long as social inequalities, historical and cultural differences, and prejudices exist, we have a host of uncontrollable variables in explaining the course of individual human lives. Our concern as citizens should be to develop a system in which talented people from any social group can become either grocery store checkout clerks or professors. Our concern as academicians should be to make sure that our fellow citizens appreciate that social barriers are principally constructs of human agency, not of nature."[42]

Anthropologist Mark Cohen adds:

> The use of IQ tests to measure the innate, genetically determined ability or intelligence of individuals, particularly individuals from different social classes, ethnic groups, or races, involves a long string of assumptions, each of which can be shown to be highly improbable if not completely spurious. To continue to measure ability for college entrance or for jobs not only perpetuates a scientific fiction; it imposes an enormous, unwarranted burden on individuals who, as a result of prior exclusion, do not participate fully in the mainstream culture. In fact it provides very dramatic 'affirmative action' for upper middle class white individuals. To use these tests without compensating for this enormous advantage, as the United States now appears intent on doing, is unconscionable, making a mockery of our protests of equal opportunity.[43]

TWIN STUDIES

Studies of identical twins are the greatest source of evidence supporting the claim that intelligence is inherited and that IQ tests can measure that inherited intelligence. Psychologists have extracted a lot of mileage from studies that looked at identical twins who were reared apart in comparison with twins who were raised in the "same" environment. Presumably the twins who grew up in the same home—same genes and same environment—provide a base point from which to compare twins who were raised in two different homes with two different families—same genes but different environment. If, for example, two twins adopted in infancy and raised many miles apart by different families still produce similar IQ test results, then it shows that intelligence is more about heredity than environment. But is this true? Can we rely on these studies to influence our conclusions about race and intelligence?

While there may be much to be learned from them, there is good reason to be skeptical about twin studies too. For one thing, just how different are the "different" environments that separated twins grow up in? Two identical twin children who are adopted by different families in America at the same time are still living in America at the same time. They may be and probably are exposed to similar burdens or benefits if their racial identification and economic class are the same. They can still be impacted in the same ways by the positive or negative biases of the teachers they encounter in childhood. Opportunities can be available or denied based on their skin color, for example, or on whether they live in Los Angeles or New York. Therefore, it is hardly conclusive proof of inherited intelligence if a white child raised in Wisconsin by a middle-class, white Christian family earns a test score that is similar to that of her identical twin sister who has been raised in Illinois by a middle-class, white Christian family. Until we see stacks of studies of identical twins raised in very different environments, say, one raised by a Swedish family in Europe and the other raised by a Yanomami family in the Amazon rain forest, it does not seem that we are justified in using twins to measure inherited intelligence.[44]

I know what you are thinking. You are wondering about that story you vaguely remember hearing something about, the one where two identical twins were reunited after a lifetime apart, only to discover that they had been living parallel lives. They ate the same foods, married women with the same name, liked the same music, and even gave their dogs the same name. Maybe you saw them on *Oprah* or read about them in *People* magazine. While such stories fascinate us and stick in our memories, they don't hold up to scientific scrutiny. Just think about it for a moment: are we really ready to believe that there is a gene that drives you to name your dog "Spot" rather than "Fido"? To prove a point, one skeptical scientist at the 1996 International Congress of Human Genetics asked attendees to raise their hands if they believed there was a dog-naming gene. Of course, no hands went up.[45] Most likely these stories, if true, are coincidences or perhaps explained in other, more reasonable ways.

WHY DO WE DO IT?

I suspect that the enduring flurry of thought and activity over intelligence testing is rooted in humans' preoccupation with ranking and outranking one another. We are so like chimps in the wild, constantly concerned with where we stand on the ladder of power. We fixate on things such as income, popularity, clothing, jewelry, cars, houses, and so on, not just to attract mates but also to claw our way up that social ladder and keep as many people below us as possible. Most people don't mean any harm, of course; it's just how we operate. It's very human. Therefore, it certainly should be no surprise that we persist, century after century, in trying to find a way to scientifically legitimize a race hierarchy. We do the same thing *within* races as well. Many English researchers tried to explain and justify an unjust class system in England as the fair result of inherited intelligence, a simple fact of nature rather than nepotism run amok. The desire to sort and rank ourselves goes deep. It is found within countries and even within families. Does it ever really end? Put two people in a room together and, assuming they are not sexually attracted to each other, their thoughts are likely to center on who is smarter and who is wealthier. It's shallow but true.

It says a lot that we still do not have a final answer to the question of intelligence and race. On one hand, the lack of convincing evidence on the matter after all this time strongly suggests that the connection between intelligence and race is a dead-end idea. On the other hand, the fact that so many people still believe in it strongly suggests that many of us feel a deep need to think of humankind as a collection of very unequal groups. Why? Is this a product of lazy thinking, an irresistible mental shortcut in our attempt to make sense of the world? Is it driven by primitive instincts? Does it come from some reptilian desire to protect territory and dominate others? Does it serve a therapeutic purpose by distracting some people from their own inadequacies and allowing them to feel superior to others and therefore better about themselves?

There are very good reasons to doubt, if not totally reject, all claims that inherited intelligence has a meaningful link to biological race. Besides the fact that biological race categories are not real, no one has shown beyond a reasonable doubt that such groups, whatever they really are, have significant differences in *inherited* intelligence. No one has convincingly argued or shown, for example, that slavery, segregation, problems within the black sub-culture, racism, and discrimination are not behind the white-black gap in IQ scores and academic achievement in general. The world has not been a fair place over the last several thousand years. It still is not and it may never be. The fact that some racial groups have fared better than others or scored higher on a test simply is not evidence enough for us to conclude that inherited intelligence is the cause of the imbalances we see between racial groups.

Anyone who looks into the idea of an race-intelligence connection will discover serious problems with it. Given the grave implications of ranking races based on intelligence, one would think that there must be a mountain of rock solid evidence that allows this idea to survive. The injustice of insisting on a race-intelligence hierarchy without overwhelming evidence to support it is a mistake of staggering proportions. It is unconscionable, cruel, and destructive. To even hint that a child—based on skin color, nose dimensions, and hair texture—is condemned to carry a burden of relatively low intelligence for life is as outrageous and immoral as it is unproven. How many children have had their confidence shattered irreparably by such claims? How many children do not try their best in school because they are aware of others' belief in the mental inferiority of their race? Expectations can influence results. Come on, some people say, there are mountains of evidence that white Europeans are more intelligent than Hispanics or Native Americans or black Africans. But where is this evidence? It does not exist. I've looked. It simply is not there. We have no mountains of evidence. What we do have are more like deep valleys of doubt concerning the claims of the racists.

Here is a typical absurdity that cannot easily be explained by those who are devout believers in IQ scores as the valid measure of something called racial intelligence. Kenya's population supposedly has an average IQ of 72. Remember that "average" would mean that many Kenyans are well below 72. So what does a score of 72 represent? Well, an IQ of "approximately 70 or below" is one of the criteria for designating a person as mentally retarded. I have been to Kenya. I have met and interacted with many Kenyans. Never once, whether I was in Nairobi or small rural villages, did I have reason to suspect for a second that Kenya is populated by millions of mentally retarded people. After my time spent there, it seems to me laughable even to imagine such a thing. To be clear, I doubt most psychologists or credible researchers would suggest that half of all Kenyans are mentally retarded. However, many racists do cite things like this national average IQ as "evidence" that black people are mentally inferior. Perhaps we should be concerned about the intelligence of people who cannot recognize how cultural differences and the lack of formal schooling with trained teachers might influence scores.

Psychologist Jefferson Fish has a relevant story from his experience with indigenous people in South America. "I lived briefly with the Krikati Indians in Brazil, who would probably score mentally retarded even on nonverbal IQ tests. Because of lack of familiarity with the materials—geometric shapes, and so on. In their village, however, I was the 'mentally retarded' one—unable to find my way around. There were no roads, [I was] unable to make and shoot a bow and arrow, unable to process poisonous manioc to make it edible, and so on."[46]

Just imagine if Krikati Indians or the Samburu of Kenya and Tanzania were the ones who were giving the intelligence tests. If they were, imagine how different the world's races would stack up based on "inherited" intelligence.

BY THE WAY, WHAT IS INTELLIGENCE?

There are many problems with the race-intelligence connection, but one of the biggest challenges is defining intelligence. That's right; there is no consensus on exactly what intelligence is. Yet this lack of a coherent and consistent definition has not stopped some people from ranking races based on average "intelligence." Think about this. Scientists do not fully understand or agree on what is meant by "intelligence," yet intelligence tests are widely believed to measure it accurately and form the basis for ranking large categories of people. Further complicating things, intelligence and race, both together and individually, are emotion-laden topics. They often stir up emotions that can cloud the science and make objectivity difficult, if not impossible. There are serious concerns, no matter how you view these issues. It's a thorny field to wander into but it must be done.

"Human intelligence is one of the most important yet controversial topics in the whole field of the human sciences," writes David J. Bartholomew. "It is not even agreed whether it can be measured or, if it can, whether it should be measured. The literature is enormous and much of it is highly partisan and, often, far from accurate."[47]

"Who knows what IQ measures?" asked the late Stephen Jay Gould. "It is a good predictor of 'success' in school, but is such success a result of intelligence, apple polishing, or the assimilation of values that the leaders of society prefer? Some psychologists get around this problem by defining intelligence as the scores attained on 'intelligence' tests. A neat trick."[48]

"The most difficult thing about investigating the genetics of intelligence is that no one can seem to agree on what actually makes up intelligence or whether standardized tests accurately measure it," writes evolutionary biologist Joseph L. Graves Jr. "In general, one can think of intelligence as the ability to reason, learn, remember, connect ideas, deduce, and create."[49]

Finally, Kenan Malik writes in *Strange Fruit*, his enlightening book about the race controversy: "If the concept of race is problematic, [then] that notion of intelligence is even more so. Intelligence has never been properly defined, no one knows what IQ tests actually measure and we have yet to identify the genes that underlie the myriad attributes that collectively give rise to intelligence."[50]

Does it sound to you as if the experts are able to accurately measure the "intelligence" of millions of people across all cultural and socioeconomic zones? Does intelligence—whatever it is—sound like something we are ready at this time to reduce to a convenient digital score? Even the men who came up with the IQ tests did not think so.

Remember, Binet and Simon may have developed the "intelligence scale" but they never meant it to become the precise measuring instrument it is thought to be today. Terman, the American psychologist, is the one who came up with the "intelligence quotient" (IQ). Binet rejected the concept of a quotient. He believed that his test was just not precise enough to justify coming up with some exact numerical score.[51] I am pretty sure that Binet would roll over in his grave if he knew that many millions of people today believe that a person with an IQ test score of 110 is not only "more intelligent" but has greater "inherited intelligence" than a person with an IQ test score of 105.

"Given the enormous range of mental abilities that all normal human beings display, the differences in 'intelligence' we measure with IQ tests surely represent only a tiny fraction of an individual's abilities, the tip of an iceberg," writes anthropologist Mark Cohen in *Race and Intelligence*. "Differences in IQ may seem large, but measured against the total of human abilities they are trivial. Therefore, to assume that genes control IQ, we have to assume that genes exert very fine-tuned control over our capabilities—far more control than genes maintain with, say, height. As a parallel example it may be possible to argue that genes contribute to the differences between someone five feet tall and someone seven feet tall; however, to assert that genes control measured IQ is the equivalent of asserting that genes control the difference between heights 5′7″ and 5′7½″."[52]

The quality and abundance of the arguments against the claims of a race-intelligence connection should tell us, at the very least, to be cautious, to make sure we do not go too far with too little evidence. There is too much at stake. University of Virginia psychologist Eric Turkheimer comments: "The theory of intelligence may be adequate, but it is far from complete, and, for the rational scientist, skepticism about its conclusions must depend, in part, on the consequences of being wrong. Seventy-five years ago, a survey of testing experts might well have concluded that immigrants from southern Europe were intellectually inferior to the resident majority [of the United States]. If it is ever documented conclusively, the genetic inferiority of a race on a trait as important as intelligence will rank with the atomic bomb as the most destructive discovery in human history. The correct conclusion is to withhold judgment."[53]

Intelligence testing remains popular despite the fact that intelligence remains a mysterious property, the fact that the degree to which intelligence is inherited remains unknown or at least controversial, and the fact that biological races are rejected by many scientists. Furthermore, despite the fact that in most cases, intelligence research subjects and test takers may assign themselves or be assigned to a race based on social rather than biological factors, intelligence testing continues to fuel the popular belief that some races have genetic advantages over others in thinking. Clearly many people are overreaching. Something as complex and elusive as intelligence is not likely to be captured and quantified by a test. Furthermore, allowing people to be included within one group or another, regardless of their real genetic makeup, destroys the credibility of the results when they are utilized in ways that suggest truths about race and inherited intelligence. There is an undeniable whiff of pseudoscience in all of this.

"The racist arguments of the nineteenth century were primarily based on craniometry, the measurement of human skulls," argued Gould. "Today, these contentions stand totally discredited. What craniometry was to the nineteenth century, intelligence testing has been to the twentieth.... We have never...had any hard data on genetically based differences in intelligence among human groups. Speculation, however, has never let data stand in its way; and when men in power need such an assertion to justify their actions, there will always be scientists available to supply it."[54]

IF THEY'RE NATURALLY SMARTER,
THEN WHY DO ASIANS HAVE TO STUDY ALL THE TIME?

I spent a lot of time in the library during my university days. One thing I observed was the large number of Asian students studying there every night. They were a minority on my campus but they seemed like the majority in the library. Many of them were friends and, looking back, I regret that I didn't infiltrate one of their study groups. Their work ethic and support structures were impressive. Their hardcore approach to studying might have taken me to another level. I was impressed with their study habits and dedication, but never thought of them as racially superior thinkers. But this is what some race-intelligence believers say today, citing the fact that Asians have been very successful academically and on standardized intelligence tests. But wait, if they are naturally superior thinkers, why do they have to study so hard? If they are so incredibly gifted, the cream of humanity's intellectual crop, even smarter than the celebrated white race, then why do they have to work so much harder than everyone else? (I should mention that although it is widely reported that Asians lead in IQ scores, there are some psychologists who dispute this.)[55] If I had consistently observed Asian students partying, drinking, playing video games, and watching TV back in the dorms and still managing to ace exams and earn straight As, then I might suspect they really are a genetically superior race. But I didn't see this. If anything, maybe it was the white, black, and Hispanic students I observed who were mentally superior. After all, many of them were able to drink heavily, do research papers the night before they were due, study while watching television, and still manage to graduate. I'm serious about this. I was a resident assistant (dorm supervisor) and had more than a few problems with drunken students hiding beer kegs in their rooms, playing soapy slip-and-slide in the hallway (don't ask), and singing loudly at 3 a.m. As I recall, all my encounters of that nature were with non-Asian students. Never once did I have to ask an Asian student to surrender the beer bong and release the terrified squirrel. Why? Probably because the Asian students were studying in the library. Psychologist Jefferson Fish's view supports my amateur observations: "In America there is a belief in individualism or something that leads people to think of some individuals as just being smarter than others because they were born with more potential," he said. "But in China, for example, the attitude is that, regardless of potential, you have to work in order to learn things. And if you don't learn it, that means you have to work harder. So my explanation for why Asians do so well on IQ tests is that they just study more."[56]

Intelligence expert Richard Nisbett agrees: "Asian intellectual accomplishment is due more to sweat than to exceptional gray matter."[57]

TOXIC ENVIRONMENTS AND INTELLIGENCE

It is vital for readers to remember an important fact about all race-intelligence associations: there is no agreement among scientists today on how much of intelligence is inheritable and how much is the direct result of environmental conditions. The childhood environment can make or break intellectual potential. It is well documented, for example, that poor nutrition; toxins in food, water, air, and soil; stress caused by a sense of danger; and insufficient attention and stimulation can affect a child's mental development. Even a child's prenatal environment is critical; the health and behavior of a mother can impact her child's cognitive development in the womb. A very long list of bad things can rob children of their potential intelligence.

Some interesting studies have shown that poor people in many societies are less likely to eat healthy foods simply because they are not available. Many urban neighborhoods in the United States, for example, do not have grocery stores that sell fresh fruits and vegetables. This can have a significant impact on the people who live there. They may lack the time, energy, or motivation to make a long journey just to buy broccoli and bananas. A "food desert," defined as a "socially distressed neighbourhood with low average home incomes and poor access to healthy food," can have severe implications for a population.[58] Because of the lack of access to healthy foods, as well as lack of money or lack of knowledge and guidance, poor children living in a "food desert" are at an increased risk of losing some of their physical and mental potential. And it's not just food. Non-white people in the United States are more likely than white people to live near landfills, toxic waste sites, polluting factories, and just about anything else that is nasty and dangerous.[59]

If many members of a particular group of people within a society are economically behind and living in unhealthy environments, it should not be very hard to recognize that the children in such a population may be losing ground every day to wealthier children who eat well, are provided with excellent medical care, and do not live next door to giant smokestacks. Given the reality of the dramatically contrasting lives that exist in virtually every society on Earth, how can anyone test children and teens from these different worlds and then hold up the results as objective proof of inherited differences?

One problem that may play a key role in sorting through all of this is the fact that "inherited" doesn't always mean "inherited" in the way race-believers think of it. An extreme example would be a child who inherits serious health problems from a mother who smoked crack during pregnancy. Yes, the difficulties were inherited, but they had nothing to do with race, as in which continent the baby's ancestors inhabited twenty thousand years ago.

One must also try to avoid the common mistake of confusing genotype with phenotype. Genotype refers to the genes of a person. Phenotype refers to the observable person. Put another way, genotype is the inherited information within a person that may or may not be outwardly expressed. Phenotype is what actually materializes before our eyes in the person, the result of genes and environment combined. Genotype is what can be or could have been; phenotype is what actually is. Genotype is the blueprint; phenotype the constructed house. Some people who believe in the race-intelligence connection are probably confusing genotype with phenotype. They see a result, such as a low score on an IQ test, and assume genotype must be to blame for it.

On matters of race and inherited intelligence, it is also important not to think of "biological" as necessarily the same thing as "genetic." Remember, in the minds of most people, races are supposed to be biological categories based primarily on genetics, having everything to do with genes and deep ancestry. But some things that are biological may not have much or anything to do with ancestry, genes, or genotype. For example, a child may possess genes that give her the potential to become a chess champion, but severe malnutrition may mean the child never becomes even a competent player, much less a great one. The same is true for general intelligence. A child who had the potential to become a Nobel laureate in physics may never master junior high school math, due to physical abuse at home or inferior teachers at school.

In his informative and inspirational book, *Intelligence and How to Get It,* psychologist Nisbett doesn't beat around the bush about where he stands: "Genes account for none of the differences in IQ between blacks and whites; measurable environmental factors plausibly account for all of it":[60]

> The evidence indicates that genes play no role.... First, all of the problems of lower-SES [socioeconomic status] people that affect ability and achievement are often exacerbated for blacks, who are overrepresented among the poor. To remind you of just what those potential problems are: They include poor prenatal care and nutrition, relative infrequency of breastfeeding, hunger, deficiency of vitamins and minerals, lead poisoning, fetal alcohol poisoning, poorer healthcare, greater exposure to asthma-causing pollution, emotional trauma, poor schools, poor neighborhoods along with the less desirable peers who come with the territory, and much moving and consequent disruption of education. For the black underclass, these problems are worse than they are for poor whites.... Additional problems also exacerbate the black situation. Black family income in 2002 was 67 percent of white family income, but black family wealth was 12 percent of white family wealth!... The unwed mother rate is 72 percent for blacks, compared to 24 percent for whites. This statistic represents a host of problems for black children, not least of which is that the poverty rate for single-parent homes is far higher than it is for two-parent homes. Perhaps equally important is the fact that such homes have only one adult, and the fewer adults there are, the less stimulating is the environment.[61]

I asked Dr. Nisbett why, given such profound environmental and cultural differences, some researchers come down on the side of biological race and genetics to explain differences in average IQ scores and educational achievement. "Since individual differences are genetic to some degree, and since environmental differences between groups just don't seem to them to be all that powerful, they infer that groups' differences are genetically based to some extent," he said. "But then they rely on dubious and indirect evidence for the claim and ignore a great deal of evidence on the environmental side. I don't quite understand that."[62]

Dr. Nisbett offers the following advice for black, Native American, Hispanic, and any other parents who want their children to achieve academic success and perform well on standardized intelligence tests such as the SAT:

Use mainstream middle-class socialization practices. And by that I don't mean mainstream *white*, I mean mainstream period. All people who have been middle class for more than one generation are inclined to do the following: Talk to kids—really talk, as in conversations, read to them, ask them questions about the world, including "known answer" questions, that is, known to the questioner, like "What color are elephants?" Ask them to evaluate things, help them to classify things, talk to them about how to understand the world, relate what they read to things in the world they already know about, relate things they encounter in the world to things they have read about, believe that their intelligence is under their control to some extent and make them believe that, demand excellence in schoolwork.[63]

Finally, I asked Dr. Nisbett to speculate about what would happen if all of America's white babies and black babies kept their genes but reversed their places in society. How might such a massive switcheroo affect average IQ scores and SAT scores? "It would reverse the score differences," said Dr. Nisbett. "Blacks would then have IQs ten points higher than whites."[64]

THE BLAME GAME

Nothing in this chapter should be interpreted as a call to excuse or shield all black Americans from any blame whatsoever for the troubling educational deficiencies that burden them disproportionately. No convincing case has been made that something called "racial genes" destined so many young black men in America to end up in prisons rather than universities. Clearly it was a terribly unfair past that set the stage for the race-based inequality we now see in America. The legacies of slavery, segregation, oppression, and widespread discrimination are real and relevant today. Racism, on both the individual and institutional level, still holds many people back. However, it seems no less obvious that there also are self-destructive elements within the black American subculture. Given the many problems imposed on them from without, one certainly does not want to blame the victims or be insensitive. But it must be said that black America is harmed when many children are encouraged to obsess over sports at the expense of schoolwork or to idolize a "thug culture" that glorifies crime. Many black Americans openly acknowledge these problems.

For example, a 2007 poll by the Pew Research Center found that 74 percent of black women and 67 percent of black men think rap music is a bad influence on black America. A majority of black people feel the entire hip-hop industry is harmful to black America.[65]

Interestingly, that same Pew poll revealed that only 53 percent of black Americans currently think of all black people in America as a single race. A surprising 37 percent of black Americans now feel that black people can no longer be described as one race.[66] This could well be a sign of growing emphasis on culture and values over traditional beliefs about biological kinship. Some people might not like this trend, if it is a trend, but I think it could be interpreted as a lean toward reality, so it might be a good thing in the long term. Races are *cultural* categories, after all, so the sooner we treat them as such, the better.

Many black American families have figured out what it takes to get within sight of the American dream. They are well aware that their cultural group was handed a terrible deal in the past and that the cards are still often stacked against them. Nevertheless, they seem to adopt an "I'll win anyway" attitude. Several years ago, while interviewing an Olympic gold medalist, I asked him how he dealt with what must have been the unsettling knowledge that some of his competitors were almost certainly taking performance-enhancing drugs. His blunt response was that he didn't care. As I recall, he said, "My attitude was to outwork them in training and then line up and beat them anyway. Let them cheat; I'll still win."

It worked for him.

DESTINY IS IN YOUR HANDS, NOT IN YOUR RACE

Despite what race-intelligence believers may say, the factors that directly contribute to the success of Asian American students are most likely parent involvement, time spent on schoolwork, and study habits. According to the National Center for Education Statistics, these are the key factors behind the success of Asian American students, not some genetic racial inheritance they lucked into. Consider the following contrasting lifestyles of student racial groups as reported by researchers:

- Asian Americans spend more time on homework than blacks, Hispanics and whites;
- Among minorities, though all groups spend time studying or doing class assignments, Asian Americans spend the most time on school assignments and studying. Other minority students spend time watching television, playing a sport or talking on the phone;
- Students whose parents were involved in their educational success knew what courses to take to become eligible for any college of their choice. However, students that were unaware of what courses to take to be eligible for college had very little to no parental involvement. Parents of the students that are unaware of what courses to take were the minority students except for the Asian American students, who experienced great parental involvement;
- The percentage of students in kindergarten through grade twelve whose parents reported taking them to a public library in the past month was higher for Asian American students (65 percent) than for white students (41 percent);
- It has been found that black students study for eight to ten hours a week, typically alone. Chinese students study for fourteen hours a week, working eight to ten hours alone and about four hours with other students, checking each other's answers and practicing their English. Their family members quiz them regularly and they work on problems kept in a public file in the library.[67]

Are we really supposed to believe it is just a coincidence that the same people who put in the most hours of study are also the people who score highest on the tests and make the best grades? Are we really supposed to believe that the academic fates of today's students were sealed more than fifty thousand years ago when humans migrated to various continents, rather than in the days, weeks, and months before the exam when students either study hard or do not? Clearly there is no need to grasp for racist conclusions, not when there are such glaring cultural and behavioral differences right before our eyes.

It is strange that environmental factors are so often downplayed or forgotten when it comes to discussions about intelligence, IQ, and race. Isn't it interesting that those groups of people who are most often accused of having inherited low intelligence from their ancestors also tend to be the same groups of people who are disproportionately burdened by toxic neighborhoods, unhealthy diets, inadequate schools, and limited or no access to good healthcare? If every generation of Native Americans over the last two hundred years had lived in America's cleanest and safest neighborhoods; attended America's best preschools, elementary schools, high schools, and universities; had no haunting knowledge of the murder, theft, and degradation inflicted upon their ancestors; consumed a healthy and adequate diet; had access to the country's best healthcare, and scored below white people on culturally neutral IQ tests, *then* maybe the race-IQ believers would be on to something. Short of that scenario, however, I just can't see a valid case for comparing the IQ test results of one "racial group" with another and drawing conclusions about inherited intelligence stretching back thousands of years.

One thing we have to ask every time the issue of intelligence and race comes up is "why?" What is the point of trying so hard to rank racial groups by intelligence? Is it to justify racism? To provide an excuse not to help fellow humans who may need it? What is the end result hoped for by those who insist on pushing this stuff? The best I've heard from the more polite race-IQ believers is that they don't want government to "waste money" on programs that are designed to improve the intelligence and academic performance of children who may be incapable of improving because of their race. The world's governments spend more than a trillion dollars per year on their armed forces and there are people who are losing sleep over the cost of some poor black or Hispanic kid getting a few more books and a better preschool teacher?

Even a majority of the most committed racists are likely to admit the obvious truth that many black people are very smart and many white people are very dumb. What, then, is the point of ranking races by IQ if so many people are then misjudged, mislabeled, and misplaced based on their race? Biologist Paul R. Ehrlich offers this useful thought experiment:

> Imagine that, contrary to everything geneticists know, there is a unitary something that could be called "genetic IQ," and some way is discovered to assess it. Perhaps someone invents a sort of smart litmus paper on which, when the paper is placed on the forehead of a test subject, a number miraculously appears, faultlessly indexing his or her "genetic IQ." Imagine further that average "genetic IQ" litmus test scores tend to

be somewhat higher in the black population, even though many whites score much higher than many blacks—some at the genius level. Would it then be good policy to give remedial aid to all whites and none to any blacks? Or would it be wiser to give additional help to those who have low scores, regardless of skin color? What, in fact, would the reason be for even bothering to calculate the group average IQ scores? Do we calculate them for populations differentiated on the basis of other characteristics, such as different blood groups? Would we want to know (or would we care) about litmus IQ average differences between those who have type AB blood and those with type O blood, between those who are tall and those who are short, or between those of normal weight and those who are overweight, or those with blue eyes and those with brown eyes?

It is only because people live in socially stratified societies and have a fascination with skin color (or nose shape or social class) that differences between certain groups are singled out for investigation and determination of their possible genetic causes. If average differences in IQ test scores were correlated with skin color in our society, would it make any sense to try to decrease the incidence of low test scores by treating skin color groups differently? Of course not, any more than it would make sense to attempt to lower the incidence of skin cancer (to which lighter-skinned people are more susceptible) by doling out sunscreen on the basis of IQ score! We should be aiding individual students with low scores regardless of skin color.[68]

Dr. Ehrlich is correct, of course. The strange obsession with races leads many people to try to link virtually everything to them. Intelligence, sex, sports, morality, music, food, healthcare, education, crime: there seems to be no end to it. We struggle to jam round pegs into square holes in order to confirm our race beliefs. We ignore the exceptions and the madness of it all so that we can maintain absurd and often harmful caricatures of humankind. It would all be laughable if so much harm did not come from it.

I once interviewed a University of Connecticut physicist, Ronald Mallett, about his work on time travel. (Yes, time travel is a respectable area of scientific inquiry these days.) Dr. Mallett shared with me the touching story about how he was drawn to the idea of time travel as boy because of the early death of his father. He dreamed of inventing a time machine so that he could travel back in time to warn his father about the heart attack he was to have. That dream led to a career in science. Dr. Mallett is a friendly and patient man, willing to gently explain complex theoretical physics to a stranger. When he excitedly described his idea of bending light and sending subatomic particles backwards in time, I could not help but admire his passion and intelligence.

I also conducted a memorable interview a few years ago with Dr. Neil deGrasse Tyson, an astrophysicist and the director of New York City's Hayden Planetarium. This brilliant and charismatic scientist did not disappoint as we talked about everything from black holes to the possible existence of extraterrestrials. During the interview, I had no doubt that I was in the presence of one my planet's brightest and most fervent thinkers.

Oh, I failed to mention that Dr. Mallett and Dr. Tyson are both identified as members of the category of American human beings known as black people or African Americans. I wonder how they might have fared on an IQ test if they had been born at a different time in American history. As for physics and astronomy, forget it. A couple of centuries ago, Mallett and Tyson might have been forbidden by law to learn to read and write! Just fifty years ago they might have been arrested, beaten, or worse for trying to attend classes at many universities in America. Today, Tyson is an academic rock star. Virtually any university would eagerly book him for a speaking engagement. Mallett's life and work may be coming to the big screen, as a prominent director has announced plans to make a film about him.

Those who are so fascinated with racial IQ averages need to explain what is to be done about the Ronald Malletts and Neil deGrasse Tysons of our species. How can society possibly sanction and refer to racial IQ averages in a way that does not discourage and squander the potential of people like Dr. Mallett and Dr. Tyson? Can we as a species really afford to discourage these minds? It's bad enough for cultures to lure, coerce, and shove people into illogical race categories at birth; it's far worse, however, to attach broad assumptions about fixed intelligence to individuals. Do those who care so much about race and intelligence care nothing about the injustice of hindering the development of brilliant minds? And this is not just about the brightest people. What about those with average and below-average minds? They too have a right to strive for their full potential, no matter what it might be. What purpose does it serve to argue—without conclusive proof, don't forget—that some races are genetically inferior? We already know the harm. What is the gain?

Given the problems facing our world in the twenty-first century, it is reckless not to value and nurture all human potential. Finding ways to care less about others certainly won't help. Moving forward, we cannot allow ourselves to be so shortsighted and inhumane that we can accept a racialized view of inherited intelligence, a position that undoubtedly ensures the loss of bright minds, some of which might have brought needed light to the world.

NOTES

1. Jefferson M. Fish, "Mixed Blood," *Psychology Today*, November/December 1995, www.psychologytoday.com/articles/pto-19951101-000038.html (accessed September 14, 2009).

2. Richard E. Nisbett, *Intelligence and How to Get It* (New York: Norton, 2009), p. 197.

3. Keim, "Poverty Goes Straight to the Brain," *Wired Science*, March 9, 2009, http://blog.wired.com/wiredscience/2009/03/poordevelopment.html (accessed September 14, 2009).

4. Jonathan Marks, "Folk Heredity," in *Race and Intelligence*, ed. Jefferson M. Fish (Mahwah, NJ: Lawrence Erlbaum, 2002), p. 107.

5. Stephen Murdoch, IQ: *A Smart History of a Failed Idea* (Hoboken, NJ: John Wiley and Sons, 2007), pp. 215–16.

6. Ibid.

7. Stephen Jay Gould, "Curveball," in *The Bell Curve Wars*, ed. Stephen Fraser (New York: Basic Books, 1995), p. 22.

8. Nisbett, *Intelligence and How to Get It*, p. 99.

9. James Flynn, interview by the author, April 13, 2009.

10. Ibid.

11. Ibid.

12. James Flynn, *Where Have All the Liberals Gone? Race, Class, and Ideals in America* (New York: Cambridge University Press, 2008), p. 99.

13. Flynn, interview.

14. Flynn, *Where Have All the Liberals Gone?* pp. 88–96.

15. Ibid., p. 92.

16. Wiliam T. Dickens, "Genetic Differences and School Readiness," *The Future of Children: School Readiness: Closing Racial and Ethnic Gaps* 15, no. 1 (Spring 2005), www.futureofchildren.org/ (accessed September 14, 2009).

17. Jefferson M. Fish, "A Scientific Approach to Understanding Race and

Intelligence," in *Race and Intelligence*, ed. Jefferson M. Fish (Mahwah, NJ: Lawrence Erlbaum, 2002), p. 20, footnote.

18. Howard Gardner, "Cracking Open the IQ Box," in *The Bell Curve Wars*, ed. Stephen Fraser (New York: Basic Books, 1995), p. 27.

19. Nisbett, *Intelligence and How to Get It*, p. 103.

20. Fish, "A Scientific Approach," p. 3.

21. Ibid.

22. Ibid., p. 5.

23. Edmund W. Gordon, "An Affluent Society's Excuses for Inequality: Developmental, Economic, and Educational," in *Race and IQ*, ed. Ashley Montagu (New York: Oxford University Press, 1999), p. 135.

24. Ibid.

25. Jefferson M. Fish, interview by the author, March 30, 2009.

26. Ibid.

27. Ibid.

28. Murdoch, *IQ*, p. 67.

29. Ibid., pp. 67–68.

30. Ibid., p. 73.

31. Ibid., pp. 74–75.

32. Ibid., p. 90.

33. Ibid., p. 143.

34. Ibid., p. 174.

35. Gould, "Curveball," p. 12.

36. To see more of the test, see MSN Encarta, "US Army Alpha Intelligence Test: Sample Questions," http://encarta.msn.com/media_461550852_761570026 _-1_1/u_s_army_alpha_intelligence_test_sample_questions.html (accessed September 14, 2009).

37. Fish, "A Scientific Approach," p. 20, footnote.

38. Ibid., p. 22.

39. Christopher Jencks and Meredith Phillips, eds., *The Black-White Test Score Gap* (Washington DC: Brookings Institution, 1998), p. 474.

40. Fish, "A Scientific Approach," p. 20.

41. Murdoch, *IQ*, pp. 220–21.

42. Marks, "Folk Heredity," p. 104.

43. Mark Nathan Cohen, "An Anthropologist Looks at 'Race' and 'IQ' Testing," in *Race and Intelligence*, ed. Jefferson M. Fish (Mahwah, NJ: Lawrence Erlbaum, 2002), p. 221.

44. Fish, "A Scientific Approach," p. 20, footnote.

45. Marks, "Folk Heredity," p. 103.

46. Fish, interview.

47. David J. Bartholomew, *Measuring Intelligence: Facts and Fallacies* (Cambridge: Cambridge University Press, 2004), p. xi, www.cambridge.org/catalogue/catalogue.asp?isbn=9780521544788&ss=fro (accessed September 14, 2009).

48. Stephen Jay Gould, "Racists' Arguments and IQ," *Natural History*, May 1974; reprinted in *Race and IQ*, ed. Ashley Montagu (New York: Oxford University Press, 1999), pp. 186–87.

49. Joseph L. Graves, *The Race Myth: Why We Pretend Race Exists in America* (New York: Dutton, 2004), p. 175.

50. Kenan Malik, *Strange Fruit: Why Both Sides Are Wrong in the Race Debate* (Oxford: Oneworld, 2008), p. 280.

51. Ashley Montagu, "The IQ Mythology," in *Race and IQ*, ed. Ashley Montagu (New York: Oxford University Press, 1999), pp. 32–33.

52. Cohen, "An Anthropologist," p. 205.

53. Eric Turkheimer, "Consensus and Controversy about IQ," *Contemporary Psychology* 35, no. 5 (1990): 429–30.

54. Gould, "Racists' Arguments and IQ," reprinted in *Race and IQ*, ed. Ashley Montagu (New York: Oxford University Press, 1999), pp. 186–87.

55. Nisbett, *Intelligence and How to Get It*, p. 154.

56. Fish, interview.

57. Nisbett, *Intelligence and How to Get It*, p. 154.

58. Kristian Larsen and Jason Gilliland, "Mapping the Evolution of 'Food Deserts' in a Canadian City: Supermarket Accessibility in London, Ontario, 1961–2005," *International Journal of Health Geographics* 17 (April 18, 2008): 16, www.ij-healthgeographics.com/content/7/1/16 (accessed September 14, 2009).

59. ScienceDaily with University of Colorado at Boulder, "Environmental Racism Study Finds Levels of Inequality Defy Simple Explanation," *ScienceDaily*, July 11, 2007, http:// www.sciencedaily.com/releases/2007/07/070709133240.htm (accessed September 14, 2009).

60. Nisbett, *Intelligence and How to Get It*, p. 118.

61. Ibid., pp. 100–101.

62. Richard Nisbett, interview by the author, April 22, 2009.

63. Ibid.

64. Ibid.

65. Pew Research Center, "Blacks See Growing Values Gap between Poor and Middle Class," *Pew Research Center: A Social & Demographic Trends Report*, November 13, 2007, p. 43, http://pewsocialtrends.org/ assets/pdf/Race.pdf (accessed September 14, 2009).

66. Ibid.

67. Grace Thomas Nickerson and William Allan Kritsonis, "An Analysis of the Factors That Impact Academic Achievement among Asian American, African

American, and Hispanic Students," *National Journal for Publishing and Mentoring Doctoral Student Research* 3, no. 1 (2006), www.nationalforum.com/Electronic%20Journal%20Volumes/Nickerson,%20Grace%20Thomas%20-%20An%20Analysis%20of%20the%20Factors.pdf (accessed September 14, 2009).

68. Paul Erlich, *Human Natures: Genes, Cultures, and the Human Prospect* (New York: Penguin Books, 2002), p. 297.

Chapter 8

RACISM AND THE WAY FORWARD

The truth is always the strongest argument.
—Sophocles[1]

I think I have no color prejudices nor caste prejudices nor creed prejudices. Indeed, I know it. . . . All that I care to know is that a man is a human being—that is enough for me; he can't be any worse.
—Mark Twain[2]

No one is born hating another person because of the color of his skin or his background or his religion. People must learn to hate, and if they can learn to hate, they can be taught to love, for love comes more naturally to the human heart than its opposite.
—Nelson Mandela[3]

People should stop fighting about their color. It doesn't make any sense. There is a time to change and I think that time is now.
—Marissa, seven-year-old[4]

It all depends on where you are born. Like for me, I don't care about race. But if I had been born in a KKK family I would care about it a lot.

—Jared, eleven-year-old[5]

That's how the world's gonna end. It's gonna be one big gang war.

—juvenile gang member[6]

"It just isn't right," he said. "Blacks and whites should get along, of course. Nobody is saying black people should be treated badly, but the races shouldn't mix like that. It's wrong."

This was an exchange between myself and someone I had known for years and had always thought of as intelligent and decent. This person disapproved of my "interracial relationship."

"But why?" I asked. "Why is it wrong?"

"It just is."

"But that's not an explanation. Why is it wrong for a white person and a black person to love one another?"

"Because they are niggers."

"Okay, end of conversation," I thought to myself. "This is going nowhere, and I feel really uncomfortable." It's as if one minute I was just having a chat with somebody and then, suddenly, I felt like some highly advanced time traveler from the future struggling to communicate with a prehistoric human who was attempting to explain tribal loyalties to me before heading off to hunt woolly mammoths. To be clear, I wasn't talking to a Grand Wizard of the Ku Klux Klan or some junior Fuehrer of the local neo-Nazi chapter. No, this was a person who walked upright, spoke fluently, was employed, and so on.

I couldn't resist politely jousting a bit more, but I just couldn't get anywhere. It is difficult to defeat "Because they are niggers" with reason. This is probably the greatest source of racism's power. It's a mountain of hate, ignorance, and blind conviction built upon a microscopic speck of actual reason and logic, if that much. But difficult is not the same as impossible. If I had felt it was impossible for humankind to overcome race belief and racism, I would not have bothered to write this book. Yes, it often looks bad, perhaps hopeless, even. For example, the United Nations held a large international conference on racism in 2009 and dozens of diplomats walked out in the middle of it because they didn't like one of the speeches. "Everyone forgot what the conference is actually about," a UN spokesperson said.[7] It's a troubling sign when world leaders come together to find ways to combat racism but are so offensive to one another that they can't even stay in the same room for the entire conference. Nonetheless, many believe it is possible to move beyond racism. Individuals manage it all the time. It remains to be seen whether we can do it collectively, but there is no compelling reason why we can't. No one should assume that we are doomed to an eternity of this irrational nonsense. In a brief 40-odd years, I myself have seen America go from a society that wouldn't accept the leadership of a black quarterback in the NFL to a country that chose to put Barack Obama in the White House. I saw Nelson Mandela win a presidential election in South Africa, and I saw a black woman travel into space aboard the space shuttle. Nothing has convinced me that race belief and racism are inevitable or permanent obstacles before us.

We made up the concept of race and we chose to believe in it. It stands to reason, therefore, that we have the ability to figure out that biological races are make-believe. We can decide to jettison race from our cultural grab bag of beliefs and traditions or at least redefine race in our minds so that it is less destructive. Racism is such a silly obsession. Can you imagine how idiotic we would look to an alien scientist from another planet who visited the Earth and tried to understand racism?

Alien: "Why does your species have this thing called racism?"

Human: "There is racism because people are different colors and have different abilities. Each race is different from all the other races. Some races are just better than other races."

Alien: "I don't understand."

Human: "It's simple. We have groups based on how the people look and their ancestry. These groups sometimes come into conflict and that's racism."

Alien: "But your groups make no sense. For example, I see many people in the dark-skinned group with lighter skin than some people in your light-skinned group. Also, I have analyzed many DNA samples and your groups are not genetically homogenous or isolated. They are not what could be thought of as subspecies. I don't

understand how you as a light-skinned human cannot accept the
dark-skinned humans as fellow members of your species with no
barriers between you."

Human: "Because they're niggers!"

Alien: "I'm leaving. Your species is nuts."

Once it has contaminated the minds of otherwise sensible people,
racism can be maddeningly stubborn and unresponsive to logic. It's understandable for people to want to give up and just accept the existence of
racism as part of the scenery, like road kill that no one will stop to clean
up. But racism is so dangerous and destructive that we should not surrender to it. We have a moral obligation to confront it everywhere and all
the time. Don't underestimate the impact of racism. This is not just about
people who use the "N" word. Racism touches the whole world and affects
millions of people. Racism is the primary reason why we need to think
deeply about the meaning of race. Racism is the reason why we have to
challenge this common belief in biological categories of human beings.
Racism turns us against one another in an elaborate global version of team
sports—but with unfair rules and a tilted playing field. Not to mention
hardship and death for the losers, in many cases. Racism helped to shape
history, and it guides the present in powerful ways. It dictates or influences
the types of people we live near, go to school with, tolerate, befriend, love,
marry, hate, fight, and kill. It also influences our decisions about which
people we help or neglect in times of need. Racism is the jagged and dangerous edge of race belief. It makes the ideas in a book like this something
more than contributions to a fringe academic debate or fodder for inane
cocktail-party chatter. Racism squanders human potential. It harms and
destroys lives. Countless millions of people have suffered and died in the
name of race. People are suffering right now because of racism, and we all
share a diminished world because of it. Due to the belief that humanity
comes packaged in individual boxes, one stacked on top of another, we are
running at well below maximum efficiency. The elimination of racism will
not usher in a utopian age. But it sure will be an improvement.

The question of whether races are real is not trivial. It is one of the
most important questions before our species today. Answering that question correctly could be the key to avoiding more horrors like those that
have plagued us over the last few centuries. Getting the concept of race

straight in our collective minds may prove to be the crucial step we need to take before advancing beyond the inward-looking tribal creatures that we are. Race belief, like the destructive aspects of nationalism, sexism, and religion, divides us and fuels irrational fears and hatreds that help to keep the world burning. Race is one of our greatest challenges. We have to reconsider our relationship with and our loyalty to what the late anthropologist Ashley Montagu called the "most dangerous myth" more than sixty years ago.

Much to my surprise, I have encountered many people who have little or no appreciation of the high cost of racism. It never fails to shock me that people hold the view that racism is something in the past, old news that is not worthy of attention. "Been there, done that; no need to dwell on it," they say. "Slavery, segregation, Holocaust, New World genocide, lynchings, and all that stuff are ancient history, end of story." Don't they know how bad it really was and still is? Have they ever really tried to imagine what it was like to be chained and packed into a cramped holding area on a slave ship? Don't they care about fairness for others today? To be honest, however, I'm not so sure I was always sufficiently appreciative of how serious racism is. Watching *Roots* on television and riding to track meets on a bus full of black kids is just not enough. I thought I had a clue, thanks to books, university courses on the civil rights movement, being exposed to a few white racists, and enduring family disapproval and a few cold stares as a white teenager for daring to date "outside my race" in the American South. But I still didn't get it. I never had a good appreciation of the dangerous consequences of racism until I glimpsed it up close and personally in the eyes of both its victims and its perpetrators. People who *really* knew racism finally enlightened me. Through them, I felt something deep in my gut. It was jarring to speak at length with people who have suffered unimaginable horrors perpetrated by those who were empowered and relieved of guilt by racism. Being surrounded by racists inside the hurricane of hatred that is a Ku Klux Klan rally is educational, I can assure you. Interviewing a calm, polite man who wants to overthrow the American government and carve out a new nation exclusively for black people is unnerving but informative. Listening to the trembling voice of a woman who survived torture at Auschwitz at the hands of a racist Nazi doctor is not something I'll soon forget. Seeing the deep sadness in the face of a beautiful young Rwandan woman who lost almost all of her family in one of history's worst genocidal spasms left me with a sense of urgency. Racism is not something we can simply hope to see improve. We must eliminate it. It was because of my encounters with those who were on the front lines of racism—in the trenches where they felt the pain firsthand—that I realized how important it is for our species to finally get it right on this issue. This is bigger than an academic debate between anthropologists and race believers. This is not about left or right, conservative or liberal, popular or politically correct. This is about what is real, and human lives hang in the balance. We must learn the truth about race and then accept that truth. If we don't, people are sure to keep on dying for a delusion.

The fatal flaws of the race concept are more than interesting lecture material for introductory anthropology classes. It is a nothing short of criminal that the basic facts of human biological diversity have not trickled down to the public. I should not have had to wait until I was a 20-year-old university student to hear about the unreality of races. This is vital information that the world needs to hear! Because we are not taught as children that races are made-up categories and that there are no such things as human subspecies, it remains far too easy for us to disconnect from our fellow humans and even hate them based on unscientific beliefs about race. I know, because I have seen the tears of those who miss their murdered loved ones.

"I WAS PREPARING MYSELF TO DIE"

I was in Kenya during the genocidal massacres in nearby Rwanda. The bloodbath claimed an estimated eight hundred thousand lives and horrified the world, but not enough for anyone to stop it, however. For months, Hutus butchered Tutsis on a scale difficult to imagine. And I do mean "butchered." Most of the violence was civilian on civilian and most of the killings were by club and machete. The Hutu-led government instigated the genocide and some military and militia elements were involved. However, much of it was as basic as neighbor killing neighbor. Every Tutsi—including women and children—was considered a legitimate and necessary target in the eyes of the Hutu killers. Just before I entered a small restaurant in Nakuru, a concerned Kenyan man warned me not to eat the fish. He explained that the fish were "dirty" because they had been feasting on thousands of Rwandan corpses that had been washed down river into

Lake Victoria. I wasn't sure if that was true or not. As I ate my pasta, I wondered why Rwandans would want to kill one another in such a personal and vicious manner. Being just a few hundred miles away from such extreme violence, occurring at that very moment, was disturbing to say the least. What was it that drove people who shared the same tiny country to such madness? I should have known. It was racism, of course.

Most press reports described the violence in Rwanda as a clash between "tribes" or "ethnic groups." But at its core it was nothing more than standard racism, the belief that groups of people are fundamentally different—based on genetics—in things such as ability and morality. Hutus and Tutsis had a long tradition of thinking of themselves as members of very different genetic categories with distinctly different ancestries and abilities. Not all Rwandans bought into this wholeheartedly, of course. But enough of them did to unleash a bloodbath of historic proportions. An outsider, say a typical white, black, or Hispanic American, would only see "black" people. To them, Rwandans would undoubtedly all be members of one race. But many Rwandans don't see it that way at all. This is where the illogical nature and pure nonsense of races as biological categories comes into play. Race is whatever a particular culture says it is. Americans see "blacks" or the "African race"; Rwandans see "Hutus" and "Tutsis." To see if I could learn more about this event, I interviewed a young Tutsi woman, Regina King.[8] She barely survived the massacres in 1994. Talking about them was an emotional experience for both of us.

"People were always made to feel different," said King. "Even from school age there were comments and prejudiced sayings that Tutsis are more beautiful than Hutus and harder working, able to get rich easier. They [Hutus] were told that they were the majority, the stronger group. From those prejudices, problems came. When I went to grade one in school, my ethnic group [Tutsi] was named in my documents. Acceptance to high school was based on your ethnic group. Tutsis became dehumanized. They [Hutus] would call Tutsis snakes. For Hutus it was like if you kill a Tutsi it is not even a sin." Virtually all Hutus and Tutsis shared the same religion: Christianity.

King said that in the southern part of Rwanda where she lived, there were many intermarriages between Hutus and Tutsis. Because of this lack of "purity," northern Hutus did not fully trust southern Hutus, according to King. "It was all very strange and evil," she added.

It is important to keep in mind that this perceived fault line or canyon between Hutu and Tutsi was not thought of as optional like membership of a club or political party. It was thought of as blood based—as in racial. Physical features were seen as important identifying traits. King explained that the Hutus tended to be tall, while Tutsis tended to have "longer noses," and "the skin tone [was] different." She says Tutsis were generally thought to be more beautiful. "But we have Hutus who are tall. We have Hutus who are very beautiful. For me it is not important."

King described stressful ordeals as a young child in school. She would have to stand up in class and be identified as a Tutsi. She recalls asking her father if she was bad because she was a Tutsi. "When Hutus from the north came in our school they were so mean and aggressive. They felt that they were superior because they believed they were pure. By grade eight I felt a lot of stress. I was at the top of my class, but I worried that I would not make it."

Believing that genetics determine a group's worth and work ethic; one group believing its members are more beautiful; ostracizing and dehumanizing people; calling people in another group "snakes" and "cockroaches"; focusing on a few physical features to exaggerate biological difference; believing in purity of blood. Doesn't this all sound familiar? King said there were warning signs but nothing that indicated how bad it would get. "I think everybody sensed it. There were threats in the newspapers and on the radio. But we did not fear genocide. We did not imagine that."

Just when the massacre was about to explode, King had plans to spend the night with friends, nothing special, just girlfriends hanging out together. But she could not go to her friends' house because of rain. All of the friends she was to spend the night with were murdered by Hutus that evening. "I was at great risk because they were looking for educated Tutsis. They wanted to kill even Tutsi women who were pregnant. They would kill the woman and take the baby out. It was so..." At this point in her story, she began to cry.

King and some Tutsi companions, including her sister, hid in the bush for a while, and eventually they found people willing to hide them in their home. But someone found out about them and told the Hutus. Soon a group of killers came and captured them. They were marched at gunpoint for some sixty kilometers to find men "who knew how to kill us in a better, more atrocious way. [Because we had hidden from them] they said we deserved a very special death," King explained. "There were different things they would do to people. They would have Tutsis set off grenades on ourselves. They would have us kill each other. They would put people in an area and then throw grenades at them." King says she witnessed these things done to people. "At that stage I was without feelings or worries. I was preparing myself to die. I was not scared."

When the group came upon a roadblock that was crowded with people, Regina and her sister managed to sneak away to safety. But Regina says she will never forget the atrocities she witnessed. She says it was so terrible that it is difficult to accept as real, even after living through it. "Husbands [in mixed marriages] were killing their wives in some areas. They were killing their nephews, their nieces."

By the end of it all, King had suffered many losses. One of her brothers was hacked to death with machetes. "I lost many other family members and friends," she said. "I lost all my uncles, many cousins, and others of my extended family."

Despite her experiences, King is not bitter or angry today. She is beautiful; her words are gentle and clear. Overall, she seems positive and optimistic. This is surprising, perhaps, in one who has been so gravely wounded by so much ugliness. Since the genocide, she has worked side by side with Hutus and not felt anger or the urge for revenge. "For me, maybe I don't know how to hate," she explains. "I don't judge anyone because of who they are. I judge them according to how they relate to me."

King even sympathizes with the Hutu killers, many of whom she believes must now be haunted by the memories of what they did. "Suffering is suffering, no matter where it comes from."

If Rwanda's genocidal bloodletting is an example of racism's evil power, then the example of Regina King is a reason for hope. She is living evidence that we have within us the potential to be smart enough and good enough to rise above the beliefs that lead us to hate and destroy based on difference.

"I REFUSED TO DIE"

Today, Eva Mozes Kor is a little old lady living in Indiana. But she is tough, as tough as anyone I have ever encountered. In Europe during World War II, she endured terrifying and agonizing medical experiments at

Auschwitz. Josef Mengele, the infamous Nazi doctor, had no concern for the life a young Jewish girl. Eva and her identical twin sister Miriam were nothing more than lab rats to him, because they were members of an unwanted race. Think about that: little girls can become expendable creatures to be tortured and put to death, thanks to race belief. Kor told me about the cruel and degrading ordeals she was forced to endure.[9] Mengele and his assistants injected her with unknown substances so that her body's reaction could be noted and analyzed. She had to stand nude in front of adults who measured, poked, and prodded her tiny body for hours. Sometimes so much blood was drawn from her that she fainted. No one ever told her what any of the procedures were for. Because sufficient distance was believed to exist between races of human beings, Mengele and others could justify these actions. The doctor eagerly selected and removed identical twins from the trains that arrived loaded with Jews who were doomed to slave labor or execution. He considered the twins invaluable to his medical research because he could experiment on one and use her genetic double for comparison. Kor had observed that whenever one twin died from his experiments, the other twin was promptly killed so that comparative autopsies could be performed. Once, Kor was near death from something she had been injected with. She overheard Mengele say that she would die within two weeks. "I refused to accept his verdict," said Kor. "I refused to die." She said she refused to die because she didn't want her sister to be killed as a result. Eva was only ten years old at the time.

Eva Kor and her sister lost their childhood at Auschwitz. They also lost their parents. So I was surprised when Kor explained to me that she had forgiven Mengele and the Nazis. She simply did not want to spend the rest of her life burdened with hatred for an individual or a group of people. Today, she owns and operates a small Holocaust museum and educational center in Terre Haute, Indiana. Sadly, Kor never completely escaped the evil that almost killed her more than fifty years ago in Europe. Her museum was firebombed in 2003. Fortunately, people—all "kinds" of people—rallied around Kor and donated enough money to enable her to rebuild the museum. True to form, Kor told me that she forgave the arsonist too. "Revenge has never cured one single human heart," she said.

"WE HAVEN'T LEARNED'"

Barbara LeDermann was a childhood friend of Anne Frank, the thoughtful and upbeat girl made famous by her diary. Their families lived in the same Amsterdam neighborhood during World War II. When the persecution of Jews became increasingly severe, Anne and her family went into hiding. Barbara, a Jewish teenager, chose to adopt a fake name and hide in plain sight. She pretended to be a Christian German girl. But she didn't play it safe: she chose to work for the Jewish underground throughout the war. She delivered food and information to Jewish families in hiding. After the war, she learned that her entire family had been murdered.[10]

Barbara immigrated to America where, she says, her life has been wonderful. Her late husband, Dr. Martin Rodbell, won the Nobel Prize for Medicine in 1994. She has four children and seven grandchildren. Still, even as she expressed sincere appreciation, optimism, and hope, I sensed great pain that has not gone away. "Just imagine six million people killed," she said. "How many of these people would have contributed valuable and wonderful things to the world? It's horrendous when you think about it. It was an enormous loss."

"Look at the people we still keep losing to violence today. That is the most frightening thing. We haven't learned. Look at what happened in Yugoslavia. Look at Africa. We still have not learned."

Most people shake their heads in disgust and feel sad about the suffering and slaughter of millions that was facilitated by the categorization of people on the basis of erroneous beliefs about their biology. But then, without a hint of reluctance, they continue to believe in, defend, and promote racial categories. Placing people in imaginary boxes does not kill them. The practice is not necessarily destructive or deadly. It is, however, extremely dangerous. We must acknowledge this danger and ask ourselves if it is worth it. When human beings cease to be human first and instead are seen as members of a subspecies with significantly different genetically determined abilities and limitations—with more or less value than others—we have set the stage for disaster. This is not speculation. It is history and it is today's headlines.

What is it about lumping large numbers of people into groups that are believed to be based on unique biology that causes so many problems? Why is it a little easier to dislike, fear, or hate someone who is identified with a race other than your own? Why does race belief so often become race destruction? I asked that question during an interview with Armin Lehman, a former member of the Hitler Youth who earned the Iron Cross for bravery in fighting the Soviet Army. He also served as Hitler's last courier inside the Berlin bunker during the final days of World War II.

Lehman described his early days in the Hitler Youth as great fun for a ten-year-old boy. He went camping, he got to wear a cool uniform, and he even got his own official Nazi knife. What more could a kid ask for in a paramilitary organization dedicated to the idolization of a creepy dictator? But there was more. Lehman says he had to sit in classrooms and listen to lectures on Aryan superiority and Jewish inferiority. He was physically examined, for example, having his head measured, in order to have his measurements added to some database that supposedly supported the case for white German racial superiority. It was in school and in Hitler Youth meetings that Lehman learned to believe in race with a passion, to hate Jews, and to feel superior to all other races. Hitler was his idol and could say or do no wrong. Lehman once saw Hitler give a speech and was swept away by the charismatic delivery. For several years, he trusted every word and every idea that came from the Fuehrer. It all fell apart in April 1944, however, when Lehman saw a frail and shattered Hitler shuffling around the bunker before shooting himself. Meeting and working directly for Hitler had been "a great honor," Lehman said, but "today I shudder when I think about it. This is a man who killed all of these people and then he treated us [Hitler Youth members] like his little kids."[11]

"There is the question of collective guilt. I never knowingly killed any of them [Holocaust victims]. I never knowingly harmed anybody." But Lehman did cause harm, of course. He admitted as much when he told me how he embraced and supported Hitler's racist worldview. In his defense, Lehman was just a ten-year-old child when he joined the Hitler Youth. He was not born with a belief in Aryan supremacy and a hatred for Jews. Devious men who were pushing a destructive racist agenda indoctrinated him into all of that. Membership in the Hitler Youth was mandatory for boys, and it was structured to be attractive to kids. He didn't seek out evil; it came looking for him. But I am not interested in defending or condemning Armin Lehman here. What is important is that his life is a demonstration of how popular leaders, governments, religions—really anyone or anything—can be completely wrong about race and of the fact that uncritically accepting their claims can be very dangerous. Just because a government agency, a school, or even your own family does and says things that suggest biological races are real and valid categories, that doesn't necessarily make it true. A little critical thinking and skepticism can go a long way toward making this a more peaceable planet.

Walter Jacobi was a member of Wernher von Braun's special weapons team during World War II. This team designed and built the V2 missiles. Slaves, victims of Aryan racism, did the hard work of building these weapons, however. This is an awkward fact of history that followed Jacobi after the war, when he and von Braun went to the United States to work for NASA. What a contradiction Jacobi's career was. Here was a man who helped design and build ballistic missiles for Hitler to use against civilian populations. Slaves worked under brutal conditions near his office. But then he went on to live a peaceful life in Huntsville, Alabama, where he made key contributions to one of the greatest human achievements in all of history. When men landed on the Moon—men Jacobi helped put there—they said they "came in peace for all mankind." Such is the flexibility of a single human life. We can destroy; we can build.[12]

Believing in biological categories and ranking them has almost always been destructive. Accepting the unreality of races—realizing that there are no real biological walls between us—at least offers the hope of building something positive. The Holocaust was not purely a racial crime. Race belief was key, of course, but there were also political, economic, and religious factors involved. One thing I took away from my interviews with Germans and Jews of that period is that focusing too much on religions, political parties, secular organizations, countries, or even races only distracts from the real problem. The Nazis didn't kill six million Jews; *people* did. The Nazi leaders who issued orders to murder civilians based on their race/religion were not demons. The men and women who carried out the orders were not monsters. These people, from the generals right down to the camp guards, were race believers who followed evil ideas to evil conclusions. How convenient it would be if they all had been monsters or at least sociopaths. Then we could simply condemn them, dismiss them, and move on. The truth is more complicated, however. The killers were every bit as human as the victims. The big difference between Dr. Mengele and Eva Kor or Barbara LeDermann is that Mengele was a devout racist with guns and muscle to back him up. The meaningful difference between

Regina King and the Hutus who tried to kill her is that she did not hate and she preferred to judge people as individuals. The greater lesson of the Holocaust and other genocidal bloodbaths is not that one form of government or another is bad or that we have to resist the grip of evil leaders. No, death camps like Auschwitz and episodes of brutality such as we saw in Rwanda teach us that typical people are capable of evil behavior. It seems to be a human trait, common to most if not all of us. Some disturbing psychological studies seem to back this up. Controversial experiments conducted by Yale psychology professor Stanley Milgram in the 1960s suggested that many, if not most, people will take orders from an authority figure and commit questionable acts, even if it means doing harm to a stranger. Stanford psychologist Philip Zimbardo conducted a famous experiment in 1971 that simulated prison conditions and revealed how easy it is for "good people" to become "bad people" when they have sufficient power and believe that questionable actions are legitimized simply by their authority.[13] There are many beliefs available that can provide sufficient motivation and escape from guilt for those who would mistreat people. Race is one of the best, however. It is so devilishly convenient in the way that it can simultaneously separate and dehumanize closely related people (white German Christians killing white German Jews; Hutus killing their own nieces and nephews; slave owners enslaving their own children). Belief in race can be so accommodating in the way it allows one group of people to feel superior and entitled based on nothing more than perceptions of a group's ancestry. It is as if we peasants were jealous and bitter because we were not born into royalty. We could never be kings and queens, so we settled for races. Better by birthright, privileged by paternity. But this can only last as long as we believe in royal blood and common blood. For race belief to continue to thrive and to endure, all we have to do is forget or ignore the common humanity within us all.

IS IT WORTH IT?

Looking back across history, scanning today's terribly unjust world, is it worth it? Does race belief give us so much that we cannot let go of it? We know the cost, or should by now—so what is the payoff? What does race belief give us that is so precious? With what do we balance torturing a 10-

year-old girl at Auschwitz or holding a gun to the head of a young girl in Rwanda? It is just not good enough to feel sad or disgusted when we hear of yet another story about some race-based atrocity. The only meaningful response is to turn away from the hollow belief that caused it. When we give our loyalty to race beliefs rather than to one another, we are all like loaded guns walking around ready to fire. The road to Auschwitz did not begin when Hitler assumed power. The Nazi death camps were the final stop for millions of people along a road that stretches far into the past. This is a road built by racism—the belief that we are so biologically different from one another that conflict is acceptable, necessary, inevitable. Sadly, the road has continued far beyond the Holocaust, which is now more than half a century behind us. Recent stops include Bosnia, Rwanda, and Sudan. Every time our minds shove someone into a mythical biological category, we help to extend that road. Every time we imprison an individual inside an imaginary box called a race, we take another step toward the next stop on that terrible road.

Cruelty is always present in war, of course, but the battles waged in the Pacific during World War II saw some of the worst race-based fury in history. Racism is not required to fuel wartime atrocities; nor is racism necessary for soldiers to hate their enemy. But it sure does make it easier. I was stunned and horrified while listening to the stories of two former American soldiers who barely survived the Bataan Death March in the Philippines. After the forced march, they endured years of beatings and starvation as slave laborers. Admittedly there were reasons other than racism for the brutal treatment of Allied prisoners by the Japanese. Many Japanese soldiers believed that it was unforgivable for a warrior to surrender, and that any soldier who surrendered was unworthy of respect or kindness. There was also a culture of general brutality within the Japanese military itself that carried over to the POWs. It was common for Japanese officers to beat their subordinates, for example. But it was racism that motivated and excused much of the cruelty. American racist attitudes toward the Japanese facilitated the savage, all-out, take-no-prisoners combat that was typical of many island battles. The Japanese, also racist, committed terrible atrocities throughout the region, largely motivated and excused by their belief that they were the natural and rightful masters of Asia. They claimed to be a superior race and believed they could therefore treat other peoples as they pleased. When the Japanese forces invaded China, for example, they executed tens of thousands of Chinese people, raped thousands of Chinese women and girls, and tortured countless Chinese civilians. Live Chinese people were bound and used for bayonet practice. For fun, some Japanese soldiers competed to see who could decapitate the most Chinese prisoners in a given time. Much of this was reported gleefully at home by the Japanese press.[14] A girl in Papua New Guinea told me that the old people sometimes talked about how the Japanese would kill local people and then hang them from trees.

In the United States, racism was behind the decision to imprison American citizens who had Japanese ancestry, including children. They were held in prison camps for the duration of the war. They were not physically abused or executed, but they were denied their rights and freedoms based on nothing more than racial affiliation. The US government did not do this to American citizens with German or Italian ancestry.

I've met and interviewed so many people who were close to key moments in World War II that I sometimes feel as if it just happened a few weeks ago. Their firsthand accounts have enlightened me about the horrors of a war that is often sanitized and glamorized by people who did not experience it. I have talked with a B29 pilot who took part in the firebombing of Tokyo in 1945. I have interviewed people close to the atomic bombing of Hiroshima, including the late Joseph Rotblatt, a key scientist who had helped invent the bomb but then became a peace activist and won the Nobel Peace Prize; the pilot of the plane that bombed Hiroshima; the man who armed the bomb during that flight; and a Japanese civilian who survived the blast on the ground that day. After talking with these people and others, I was left with a sense that the war in the Pacific was something more than a clash of empires. A marine who fought on the island of Peleliu told me about the vicious combat between the Japanese and Americans. He described how flamethrowers were used to "burn out" Japanese soldiers from caves. There was virtually no mercy or restraint from either side. The level of hatred and violence in the Pacific theater seemed to go far beyond the level in Europe between American/British and German/Italian soldiers who were mostly white people fighting white people. Interestingly, the fighting between the Soviet Army and the Germans on the Eastern Front was far bloodier and savage than the fighting on the Western Front, more like the fighting in the Pacific. Perhaps that is because it too was viewed as a clash between different races. According to Hitler, the Eastern

Front represented an epic fight to the death between Aryan and Slavic *races.*

Overall, the war in the Pacific was amplified and intensified by racism. Both sides believed it was something more than nation against nation or this army against that army. It was a war between different races—different kinds of people. Hector, a US Navy veteran, survived torpedo strikes and kamikaze attacks onboard the USS *Intrepid,* an aircraft carrier that saw plenty of action in the Pacific. He is a calm and friendly man but becomes noticeably uneasy when he talks about his former enemy. "I hate to say this but when I see a Japanese person it gives me a bad feeling." He explained to me that even long after the war, if he saw a Japanese person he would "see a kamikaze." It was so bad that he had to "see a psychiatrist in order to calm down." It was my sense that Hector was a good man who knew better than to hate Japanese people sixty years after the war, especially Japanese people whom he had never met. But race belief seems to have become entangled with his war experiences and to have stolen his peace of mind forever.[15]

I interviewed several white American veterans of World War II who took part in the D-Day invasion and fought the Germans across Europe. While many of them were haunted in one way or another by the war, none of them said anything about harboring anxiety or lingering anger toward white people with German ancestry.

I interviewed two veterans of the Pearl Harbor attack. One of them holds a grudge against all Japanese people to this day. The other one, Richard Fiske, however, went so far in the other direction that he actually became a very close friend of some of those who once bombed him. Fiske, a US Marine bugler turned rifleman, survived both the sinking of his ship at Pearl Harbor and the bloody invasion of Iwo Jima. Many years after the war, despite suffering severe posttraumatic stress syndrome, he became a close friend of Zinji Abe, one of the Japanese pilots who had attacked Pearl Harbor on December 7, 1941. "Abe and I met and were just drawn together for some reason," said Fiske. "I've come to know other Japanese pilots too." In the 1990s, while Abe was visiting Hawaii, where Fiske settled down to live after the war, "Abe and I were sitting at the hotel before he had to leave to go back home to Japan. He put his arm around me and said, 'Richard San, please, I want you to do me this special favor. Here is $300; please buy two roses every month, one for me and one for you. Take them to the *Arizona* [a US battleship sunk in the attack and now a memorial]. This is my simple way of saying I am so sorry.'"

"I hope that people will look at Abe and I and use us an example for world peace and reconciliation," said Fiske. "We were vicious enemies at one time and now we are the closest of friends. I live by three things: friendship, love, and truth. I wish everyone would.... I've seen enough death. I don't want to see anymore. By God, it's so simple to sit down and talk. Why don't people understand that?"[16]

That is a touching story but not such a rare one. Many veterans of many wars have found friendship with former enemies. The important point about a story like Abe and Fiske's is that the genetic makeup and ancestry of these men did not change after the war. But while their racial identities did not change, their hearts certainly did. They were able to finally see through that imposing wall of race between them—just as if it had never been there.

In 2009, former Ku Klux Klan member Elwin Wilson apologized to US Congressman John Lewis, a black man, for taking part in an attack on him nearly half a century ago. Lewis accepted the apology and said it was another example of how love can overcome hate.[17] Clearly even the most intense prejudice can be defeated. It is possible. Isn't this compelling evidence that racial conflicts are not inevitable, hardwired, or "in the blood"? If some individuals can reverse course and abandon racism, can everyone, can all of society?

Dr. Nick Wynne, a southern historian and a former director of the Florida Historical Society, grew up in rural Georgia where he saw segregated bathrooms, water fountains, and waiting rooms at the bus and train stations in his town.

"As a native Southerner, I was exposed to the racism inherent in the Jim Crow system of segregation," he said.

> Although as a young person I am sure I made use of racist language, my mother and father tended to be more judgmental in terms of achievement than they were of race. That is, they tended to think what an individual did as being more important than skin color. Economically we were poor, but we were never poor in encouragement to become achievers, to gain an education, and to think independently. The end result was that the familial milieu I experienced was less racist and more focused on accomplishments. My two brothers and I left the small rural environments of our youth early on and secured college educations. My

sisters who remained in our hometown did not go to college and married into local families. Although we siblings were raised to think for ourselves, my sisters became more accepting of the prevailing racism that was cloaked in the guise of religion. Interestingly, their idea on race, while putting whites in a superior category, is more theoretical and is never practiced on a personal level. That is, they use the "N" word, believe whites are superior to other races, yet they interact positively on a one-to-one basis with members of all races. There appears to be a great deal of cognitive dissonance at work here.[18]

Dr. Wynne is guardedly optimistic about human progress:

Like the rest of the world, the United States has made tremendous strides in realizing that race should be the least important gauge for evaluating human relations. Scientific developments, particularly regarding the sameness of human physiology and intellectual capacity, have been instrumental in obliterating perceived differences. Tragically for the United States and other countries, modern religion, which theoretically embraces the concept of universal acceptance, has seemingly reverted to the prejudices of earlier centuries and tends to be more unaccepting of reality. Critical to the "new" theology is the rejection of Darwinian ideas of evolution and change. Once located primarily in the religious South, the attacks on Darwinism now seem to be widespread among the fundamentalist communities across the United States, aided and spread by the ultraconservative branches of the Republican Party. In many ways, there is an intellectual bridge that joins these political-religious conservatives with the more extreme elements of white society, such as the Aryan Brotherhood, the Ku Klux Klan, and even more radical groups. Although some of the more educated conservatives may dress their opinions in the rhetoric of "family values," the reality is that their views are little different from the openly racist views of a hundred years ago.

"If the concept of race could be abandoned, it should be," Dr. Wynne added. "However, one wonders if humans would not find some additional justification to declare and maintain their beliefs in differences."[19]

THE GUARDIANS OF HATE

Brooksville, Florida, presented the ideal setting for a public Ku Klux Klan rally. Not because there's an abundance of hate in this attractive town with a population of less than 10,000—I have no reason to suspect that—but because the scenery and acoustics are perfect for an afternoon of raw racism. The hot sun feels about right, and the tall marble statue of a Confederate soldier on the grounds of the courthouse is a nice touch. There is also Brooksville's quiet and sleepy atmosphere, which amplifies screams of "white power." These shouts seem much louder here than they might seem in New York or some other big city. The town's history works, too. Brooksville was named after white South Carolina congressman Preston Brooks, the man who severely beat Senator Charles Sumner of Massachusetts with a cane on the US Senate floor in 1856. Sumner, a staunch abolitionist, had delivered an antislavery speech that angered Brooks. During the tense and divisive times leading up to the Civil War, Brooks's attack quickly became legendary in the South. Brooks rose to rock star status among southerners, and naming a town after him was the least Florida could do to honor him. There is also the disturbing claim that Hernando County, in which Brooksville is situated, had the highest recorded rate of lynchings per capita in the entire United States in the early twentieth century.[20] Add to all this a phalanx of Ku Klux Klan members standing on the courthouse steps, an army of police dressed in full riot gear, a crowd of onlookers made up of neo-Nazis, antiracism protestors, and angry black people, and you have the makings of an interesting afternoon. My only previous encounter with the KKK was during a summer vacation in West Virginia when I was a young child. While exploring the woods alone, I stumbled upon a giant Christian cross in a clearing. It was black, charred by fire. No one was around, but I was scared. I sensed hate all around me.

Staring racism directly in the eye is very different from reading about it or watching fictionalized depictions of it on television or in movies. Feeling the heat of racial hate up close and personal is an eye-opener. I highly recommend it for everyone. It's certainly an effective antidote to apathy on the issue of racism.

I'm scared again this strange day in Brooksville, just as I was when I was a little boy back in West Virginia staring up that charred cross in the woods. I'm not scared about the future of Florida, America, the world, or humanity. I haven't had time to think about all that. Right now I'm just scared about my own safety. I understood, coming here, that I wasn't getting on a ride at Disney World. One expects hate and ugliness to show up at a Klan rally, but I'm still shocked by what is in front of my eyes. If America is infected with racism, then this is one of the open wounds. And it's nasty.

"White power!" screams the speaker. This slender man with orange hair seems to be enjoying the moment. His eyes widen every time he delivers a line. His talking points are powerful—ignorant and mean, yes, but very effective in stirring up the crowd. Sweat drips off his face as he digs in and begins to work hard to make the case that white people are the rightful rulers of the universe. He wears a white Klan robe but no hood. He whips his bullhorn around and aims it at the crowd as if it's a pistol. Several Klansmen stand behind him at parade rest like an honor guard.

"White Power! We have to take back America from the Jews and niggers* today!" he yells. "This is a white nation for white people!"

He continues delivering more racist clichés, all lines that have been said countless times before by countless racists. But while there is nothing fresh or particularly interesting in the words, his delivery is above average. He screams a line or two, then pauses in order to give the KKK supporters a chance to chant back in support. I'm sure he wouldn't appreciate the comparison, but he reminds me of traditional Southern black preachers with their rhythmic speech patterns and regular pauses to allow for "amens" and words of reinforcement from the congregation.

"White power!" he screams. "We can't keep letting niggers take our jobs and move into our neighborhoods. America is a white nation and we have to keep it white for our children!"

These are scary words. Holocausts and Crusades sometimes break out after such speeches. It is the audience around me that I worry about now, however. The speaker and the rest of the KKK contingent are a safe fifteen yards away from me on the steps of the courthouse. A sizable police force stands between them and the audience. But there is nothing standing between that audience and me. Of the approximately two hundred people present, some are Klan supporters and others are younger Nazi/skinhead

---

*Unlike some other offensive words in this book, the word "nigger" has not been censored. This is because of its direct relevance to the topic of racism. Readers should not interpret this inconsistency as any endorsement of the usage of that word or a lack of awareness about its power to offend.

types. Some very angry black people are also in the crowd. There are also several black and white people standing together and chanting peace and love slogans. Clearly, they are not locals.

A large block of black people begin repeating, "F— the KKK!"

Suddenly a black woman confronts one particularly vocal Klan fan in the crowd who had been yelling "white power" with great enthusiasm. She faces him and with both hands frantically points at her crotch and invites him graphically to use his mouth for something other than shouting.

Despite the vulgarity of her words, the woman does not appear to be out of control or ready to commit violence. It seems to me that she is doing nothing more than toying with the white man, not really taking him seriously. The poor racist, I almost feel sorry for him. He tries his best to maintain his angry demeanor, but he just can't do it. Her taunt is too far over-the-top and he cracks up laughing. Several black people around me stop chanting "F— the KKK!" and laugh, too. I am still scared but I laugh, too. Wouldn't it be great, I thought, if every racist blowup could end on such a lighthearted note? Maybe the fact that so many of these irate demonstrators are able to burst into spontaneous laughter is an indication that this racism stuff doesn't always run as deep as we assume. When the two sides lose their composure and start laughing, they suddenly look like actors who have been playing roles in a movie but have lost their concentration and broken out of character. I start to wonder. Are these people just playing roles? Is this all an act, a big production? Are they just puppets in a play? I don't doubt their sincerity in hating their fellow man, but I suspect this is more about following a script than anything else.

Unfortunately, the show must go on. The angry screaming resumes. As entertaining as the light-hearted laughter had been, thirty minutes later I begin to feel more nervous than ever. Unlike most of those around me, I suspect, I'm a serious student of the civil rights movement. I have read books about it, watched documentaries, and taken university classes on the subject. This scene feels way too familiar. I think to myself: "Yeah, it's all fun and games until the police start shooting tear gas at us." It always starts with the speeches and chants, but later there's running, screaming, and concussions."

I can feel the tension rising in the crowd. It's getting out of hand. I see the police responding to the perceived escalation by shuffling men into different positions. Meanwhile, the KKK speaker is relentless with his cruel and inflammatory words. His pace hasn't changed from his first sentence. He's got stamina; I'll give him that. I sense that some of the black people in the crowd are getting close to their daily tolerance limit for hearing the word "nigger" yelled at them by white people. With the danger increasing, I begin to assess my situation and plan the best escape route if things become violent.

It doesn't look good for me. Where does a raceless human hide at a Klan rally? My physical appearance is similar to that of the KKK and neo-Nazis but I can't seek refuge with them because I'm pretty sure they won't accept me. I can't run for sanctuary among the black people because, to them, I probably look a little too much like the men who have just spent the last hour threatening and berating them. I can't even consider seeking sanctuary with the tie-dyed peace activists. These peaceniks probably won't last thirty seconds if a riot ignites. I feel very much alone.

Then, thanks to a dramatic event unfolding on the courthouse steps, I forget all about my impending doom. One of the Klansmen who has been standing behind the speaker suddenly throws down his hood and takes off his white robe. It's difficult to hear him but I make out: "I don't want to be part of this. I'm for white people; I ain't against black people."

Well, that's interesting. He didn't know the Ku Klux Klan is against black people? This goes to show that one should always research a hate group thoroughly before joining it. The man storms off the steps toward a side street. Instinctively I dash off to intercept him for an interview. I have a million questions to ask him. I want to know why he has had the change of heart, what attracted him to the KKK in the first place, and would he kiss Halle Berry if she and he were the last two humans on Earth. No chance. I make it to within twenty yards of the man, only to be stopped cold by what appears to be a very large knight holding a wooden utility pole.

"Halt!" screams the policeman. He thrusts a massive wooden club at my abdomen. It doesn't hit me with force, but I feel it and know instantly that I don't want that thing smashing my bones. He's wearing a huge bullet-proof vest that seems excessive, like the protective gear worn by bomb squad guys. His black gloves, helmet, and visor intimidate me further. I'm not used to being yelled at and confronted by the police, much less by a policeman who is dressed for medieval combat. Before I have time to explain to him that I am not a racist protestor or an antiracist protester or any other kind of protester, he screams: "Halt!" I want to say that I'm just a neutral observer, a curious civilian, a writer, a harmless fly on the wall, a native Floridian; I come in peace....

"Halt!"

I want to tell him that all I'm trying to do is find out why our species is filled with so much self-hatred in the form of racism....

"Halt!"

Clearly this police officer is not willing to deviate from his programming. I suppose he thinks I'm trying to attack the man for leaving the Klan, or maybe attack him for having been in the Klan. I sympathize with the officer. It's not easy being a policeman on good days. When an entire town is on the verge of a race riot, it must be extraordinarily stressful. Given the tense atmosphere, I completely understand why he is confronting me like this, but I try to explain one more time.

"Sir—"

"Halt!"

"Why in the hell does he keep screaming that over and over?" I think to myself. "If I 'halted' anymore my cells would be collapsing inward. I'm not threatening anyone, so why is he so aggressive? I'm unaffiliated. I never chanted all day, not even once. If I said the 'N' word in anger once, I would spend the next ten years in therapy trying to deal with the guilt. I'm not a part of this."

"Halt!"

This time he screams in a way I interpret to mean that the next voice I hear will be that of a paramedic asking me if I can wiggle my toes. Still, I want to get to that ex-KKK member. Then I feel the policeman's club press into my stomach. I decide that's my cue. It's time to seek safety back in the crowd of angry screaming racists.

"We want niggers to go back to Africa! It's best for them and it's best for America!" shouts the KKK speaker. "Niggers belong in jungles, not in the USA! This country was built by white people, for white people! Look at our schools! Look at the crime in our streets! We have to reclaim America and make it a country for decent white people again. It's obvious that the jungle bunnies can't compete with white people. If they could, why would they need affirmative action and welfare?"

Cheers and jeers from the crowd punctuate every statement. The young neo-Nazi/skinhead types laugh and clap. Some black people boo. The loudest black people continue to chant "F— the KKK!" For a moment, it feels very much like a high school football game. Then, for another moment, it's like a really low-budget Nazi rally from the 1930s—plenty of passion but none of that Leni Riefenstahl flair.

The rollercoaster ride picks up speed. I feel energized by the high drama and passion but then promptly drained by the depressing depravity of it all. Mostly, however, I feel lost. It's not as if I'm an alien from another world. I was born and raised in Florida. Brooksville is not my hometown, but it's close enough. I know these people. The black people, the white people, they are all familiar to me. Okay, the antiracism protesters seem a bit odd. Apart from them, however, I don't feel like these are strangers all around me. These are the same sort of folks I once sat with in elementary school classes, played Little League with, and passed in the aisles of the grocery store. But while I may not feel like a stranger, I do have strange feelings. I feel disconnected from this entire scene. Why, I wonder, don't I need to belong to a race and view the world through the prism of race? It seems such an obvious and irresistible force in society, especially in the middle of this crowd today. In high school, why was I able to draw inspiration from great African runners without ever feeling that we came from different worlds or belonged to different subspecies? Why don't I feel biologically distant from nonwhite people? Never once throughout my travels in Africa, South America, Asia, the Middle East, and the Pacific islands have I felt that I was in the presence of someone who was fundamentally biologically apart from me. Yes, I have encountered people who spoke, dressed, ate, and thought differently from me, but I always knew that it was culture not DNA that put those differences between us. I knew that if I had been born where they were born I would probably be speaking, dressing, eating, and thinking just like them. The crowd keeps screaming. I glance over at a teenager with a large white swastika on his black T-shirt and wonder why it is so easy for me to live without race belief. Am I mutant? Was I born without the race-belief gene? Did my parents somehow fail to properly indoctrinate me into the religion of race? Or was it the anthropology classes that saved me?

I understand that it's probably easier for a white person to live off the race grid and think of herself or himself as raceless. Surely it is harder to forget or reject race when you feel the direct, negative effects of racism every day. Had I been born a Native American, black, or a Jew in the Middle East, it would probably be more difficult for me to live a raceless life. I feel fortunate to have escaped much if not all of this madness. I wish others could escape it. Standing here—in the middle of all this race belief run amok and the hate it has generated—makes me grateful to be me.

A few black people in the crowd start yelling direct threats at nearby white racists. "I'll kick your f—ing ass, cracker!" screams one. "Go back to Africa, you spear chuckers!" yells a white man who is in desperate need of a racial slur update. ("Spear chucker" is so 1970s.) Then, oddly, the chanting from the two loudest groups becomes strangely coordinated.

"White power!"
"Black power!"
"White power!"
"Black power!"
"White power!"
"Black power!"

They may hate each other, but whites and blacks can certainly cooperate when it comes to their protest chants. My mood begins to sour. I am reluctant to be judgmental, but this foul smoke of hate and ignorance that is twisting all around me is becoming annoying. I want to throw up. I always try to understand and sympathize with people who think differently from me. I know that everyone arrives at the present from different backgrounds and experiences. As an anthropology student, I was trained to resist the pull of ethnocentrism and work hard not to become a condescending ass. As a Star Trek fan in childhood, I was trained "to explore strange new worlds, to seek out new life and new civilizations." But now, hearing so many angry and misguided misfits screaming like this, it's hard to be humble. If these people are the products of race belief—something anthropologists reject as unreal, remember—then we need to hit the reset button and make some changes in America and throughout the world.

"White power!"
"Black power!"
"White power!"
"Black power!"

What else is this but stupidity in stereo? Such a sad display of hate and twisted pride would be just a silly sideshow if not for the depth of obsession it reflects. A Klan rally is a symptom of a country and a world that has serious mental problems. Today, the Ku Klux Klan is well past its glory days but not near extinction either. A 2009 study by the Southern Poverty Law

Center found that hate groups have boomed in the twenty-first century. Since the year 2000, there has been a 54 percent increase in the number of hate groups in the United States. The Southern Poverty Law Center attributes this to fears about immigration, economic problems, and the election of President Barack Obama.[21] It seems likely that the Klan and other such racist groups will live on as long as race belief lives on. They will always dwell somewhere on the fringe, at least, as the most disturbing symbols of racism. Even if the Klan finally dies one day, I wouldn't be surprised if it promptly returned from the grave to stagger around like a starved zombie, clawing at human decency wherever it finds it. As long as there is deep and widespread faith in the existence and importance of biological races, there will be a home for the KKK. But the Klan is not the problem. Men in robes yelling "white power" are just providing window dressing. Race belief and its inevitable byproduct of racism are the problems.

WELCOME TO THE REVOLUTION

While driving through the "black section" of St. Petersburg, Florida, I noticed that many buildings were old and worn. Some appeared to have been burned and left to rot long ago. The place looked like rubble left behind after some war that nobody ever bothered to name. An old man in tattered clothes was walking on the side of the road but barely seemed to be making forward progress. He didn't just look old; he looked tired. A shirtless child stared at me as I slowly drove by. He looked at me as if I were from another world. Perhaps I am.

I had come to this neighborhood to meet with members of the African People's Socialist Party, an organization that wants to secede from the United States "by any means necessary." Hey, if a bunch of white people in South Carolina could do it, why not some black people in Florida? One of my white friends thought I was crazy to come here alone. But after speaking to one of the party's leaders on the phone, I felt confident that everything would be fine. I'm just not afraid of black people; plus I was confident that they would sense my inner goodness and refrain from killing me.

I found the building without difficulty. There was a big sign out front, announcing precisely who was inside and what the group was about. They may be a revolutionary movement with goals that could be defined as treasonous, but they weren't being secretive about it. I walked inside and was met by two men in a lobby area. One smiled and said hello. The other did not. He was cold and looked angry. I leaned toward the nice one.

I told them that I was there to learn about their organization. They led me down a hallway. The one who was unfriendly had not rattled me too much, but the posters on the walls did. They virtually screamed at me as I walked by.

"Racism! Oppression! Lynchings! Thieves! White murderers!"

One poster included a large black and white photo of a black man hanging from a tree. White people were all around him. Some of them were celebrating. I suddenly felt very uncomfortable. It was as if a few centuries of white-black hate had suddenly been dumped in my lap. Then again, maybe it had always been there and I just hadn't been paying close enough attention.

Finally, we reached the leader's office. It was a small, barren room. He rose from behind a desk and greeted me like a friend. He certainly did not react to me as if I was a member of the evil empire that was his sworn enemy. That was a relief. Another man was in the room with us. He wore a tight white T-shirt and blue jeans. He was huge; about a nine on the intimidation meter. He sat there silently. No one introduced me to him, and I sensed he didn't have any interest in knowing my name, so I kept my attention on the leader. The men who had walked me in left without saying a word.

I had done my research and knew all about the African People's Socialist Party. Back in 1966, it grew out of JOMO (Junta of Militant Organizations), the Black Studies Group in Gainesville, and the Black Rights Fighters in Fort Myers, Florida. Heavily influenced by the California-based Black Panthers in the 1960s, this group was looking for a new society, one without white people. My first question was not gentle. I asked why they believed anger, separation, and racism were the proper responses to problems caused by anger, separation, and racism. I also wanted to know why the group seemed to feel that violence would be an acceptable way to achieve its goals.

"We have to consider violence as the way to improving the situation of black people in America," the party leader said. "They [white people] have all the power, money and resources. We cannot imagine them giving it up peacefully."

That's a disturbing belief: injustice and racism will reign forever unless violence stops it. I certainly hope that's not true. Personally, I can't imagine how racism can cure racism. It seems like the wrong path to me. From across the small table, I sensed a cold arrogance within the leader. Not that he wasn't nice. He was very polite, but there was something about his posture and tone that disappointed me. He reminded me of someone I had once listened to on a sunny afternoon in Brooksville, Florida. No, he wasn't screaming vile threats and insults, but there was still an ugliness tainting this otherwise handsome man.

He was so sure of his position, confident that his kind of people must separate from other kinds of people. Clearly he had a valid point about the historical injustice, brutality, and crimes committed against black Americans, but I couldn't overlook the fact that, in terms of his ultimate hope, he was not so far away from that Klansman with the bullhorn back in Brooksville. He too felt embattled and exploited. He too claimed to be a victim of racial injustice, sold out by Jews and traitorous whites. He too wanted his own racially pure nation.

The man continued, explaining to me that his organization believes black Americans are a colonized people, still being exploited by America, the "mother country." I recalled a chilling line from *Soul on Ice*: "We shall have our manhood. We shall have it or the earth will be leveled by our attempts to gain it."[22]

"We want our own government," said the leader, "our own businesses, our own lands."

"It seems like you view all white people as one unified group," I said. "Do you think that all white people today are responsible for injustices blacks have suffered and continue to suffer? Do you hate all white people?"

"No," he replied. "We don't hate all white people. But most are accomplices in some way to what is going on. They at least condone it."

"What about white people who sincerely want peace and justice for blacks and everybody else?" I asked.

"We don't want or need their help," he said. "This is a black crisis and we must find a black solution."

"So what is the 'black solution'?" I asked.

"Self-determination," he answered. "We want nothing to do with white America."

Judging by their official wish list, they mean it. The African People's Socialist Party demands billions in reparations from the federal government and wants all blacks to be released from jails and prisons. The party views them as political prisoners. It also demands full ownership of five southern states for black people to live in.

The party literature describes America as a nation "founded on the genocide of native people, the theft of their land, and the forcible dispersal, enslavement, and colonization of millions of African people." While I can't dispute those facts, I reject the idea that violence and separation are the best possible answers to that terrible past.

These people are no visionaries. They just want some payback. The white man got his, and now they want to get theirs. Sadly, their core philosophy mirrors the system that bred the crimes they detest. They seek to become what they hate. But I kept those thoughts to myself as I smiled and said goodbye to the black rebels of St. Petersburg. Although I view his goals as destructive and vulgar, the man I spoke with that day was pleasant and thoughtful. We might easily have become friends—if I hadn't been on the other side of his imaginary wall, of course.

It is long past time for the public to hear and accept what the scientists are saying. If race is a cultural lie masquerading as a biological truth, then everyone needs to hear that and accept it. The idea of racial categories has provided humankind with the structure, motivation, and all the absolution needed to freely prey upon itself. The cost is too high to look the other way. Too much blood has been spilled for us to avoid challenging the belief. It took me a while, but even in the foul speech of the Klansman and the angry words of a black separatist, I was able to see through the hate and discover something positive. In their words of hate, destruction, and separation I found unity and hope. For the guardians of hate may come in different colors, but even they, like all of us, are more alike than different.

THE RACIST WITHIN

I am happily married to a beautiful woman who would be identified as "black" by most Americans. I have logged far more time around black people than the average white guy. I've been to Africa twice. And I listen to the Miles Davis masterpiece, *Kind of Blue*, at least twice per month. Much to my surprise, however, these impressive credentials don't mean a whole lot. I'm still a racist. Well, maybe it's not quite that bad, but something seems to be creeping around in my subconscious that shouldn't be there. I discovered my inner racist by taking the IAT (Implicit Association Test)[17] that is offered online to the public as part of an ongoing study by researchers at Harvard University, the University of Virginia, and the University of Washington.[23] The test has the test taker react as quickly as possible to images of black and white faces, linking them alternatively to positive and negative words such as "hurt," "awful," "pleasure," and "good." It seemed like fun until I saw my result.

I suppose I can take comfort in the knowledge that I am not alone. Of the more than a million people who have taken the test to date, 70 percent show some automatic preference for white people compared to black people. About 12 percent showed an automatic preference for black people over white people. This adds up to 82 percent of test takers with a race bias in one direction or the other. Sadly, only 17 percent of participants recorded "little to no" automatic preference for white people or black people. We should find these people, let them run the world, and encourage them to have as many children as possible.

I was relieved to read on the test Web site that about half of the black people who have participated also show a preference for white people. This is a strong indication that the source of the bias is almost surely cultural. It does not support the existence of some innate evil within us, telling us to shun and despise people of other races. If that were so, then black test results should roughly match white test results in bias for their race and against the other. If there really are cultural inputs that influence us with negative images and messages about black people, then black people probably would be vulnerable to this influence as well to some degree. And apparently they are, according to the IAT results.[24]

Once upon a time one might have thought that watching television news was a good thing. Maybe not, however. Two 2008 studies found that people who watched more local or network news in America were more likely to see black people as intimidating, violent, or poor. Researchers found that, based on police crime statistics, American news coverage overrepresents blacks as criminals and whites as victims.[25] Apparently all those news reports showing black men in handcuffs matter. Maybe the thousands of music videos showing diamond-laden black thugs with a gun in one hand and a gyrating vixen in the other have consequences after all. Maybe decades of questionable casting by Hollywood and television producers have left their mark on the minds of the masses. White is good and black is bad. It's harsh and it's not fair. But it's out there, and apparently it creeps in no matter how fair and just you want to be.

I can't help but shake my head at the profound irony of white-is-better-than-black bias. By the tens of millions, black people have been enslaved, exploited, murdered, and discriminated against over the last few centuries. Meanwhile, the white race has genocide in the New World, not one but *two* world wars, and the Holocaust on its resumé. But look who gets labeled "bad," "violent," and "dangerous." Now *that* is effective marketing.

I don't want to be a racist, on the surface, deep down, or anywhere else. So my immediate reaction to my IAT result was to explain it away as fast as possible. Perhaps it has something to do with the fact that I frequently write about poverty, something that disproportionately affects black people in the United States and worldwide. Perhaps my heightened awareness of the suffering and injustices experienced by many sub–Saharan black people, for example, leads me to associate "hurt" or "bad" with a black face faster than I do with a white face. If so, this is not a bad thing. It might be a sign of compassion. I'm not sure about this, but it's my excuse and I'm sticking to it. Sadly, however, it is all too possible that growing up in the United States and living my life in a race-crazy world has left me mildly infected by the race virus.

Fortunately for me and the more than 80 percent of IAT participants who also seem to have an implicit racial bias, there is no reason to think it has to determine our behavior in the real world. An implicit bias in favor of one race or another does not mean that one is necessarily an overt racist or will treat people who are identified with other races poorly in daily life. The IAT site offers this note for test takers with a score that indicates automatic bias:

> Social psychologists use the word 'prejudiced' to describe people who endorse or approve of negative attitudes and discriminatory behavior toward various out-groups. Many people who show automatic White preference on the Black-White attitude IAT are not prejudiced by this definition. It is possible to show biases on the IAT that are not consciously endorsed, or are even contradictory to intentional attitudes and beliefs. People who hold egalitarian conscious attitudes in the face of automatic White preferences may be able to function in non-prejudiced fashion partly by making active efforts to prevent their automatic White

preference from producing discriminatory behavior. However, when they relax these active efforts, these non-prejudiced people may be likely to show discrimination in thought or behavior.[26]

So the bottom line from the IAT researchers is that most people have a racial bias but it doesn't mean that they have to be racists. Regardless of whatever internalized unfairness we may carry, we can think through it and choose not to be prejudiced individuals. We just need to remember to keep our guard up. On the other hand, we cannot dismiss the likelihood that widespread implicit racial biases drive a lot of the injustice we see. With regard to the race factor, President Barack Obama's 2008 presidential campaign went a lot more smoothly than most people probably would have predicted. But there were many disturbing moments along the way. "Obama-monkey" T-shirts were sold during the campaign. Some people even showed up at Republican rallies with "monkey dolls."[27] Then, in early 2009, the *New York Post* published a stunningly insensitive cartoon that depicted two policemen shooting an ape; the caption referred to President Obama. A few people waving ape dolls at a political meeting and a cartoon do not threaten civilization. But such things influence other people, even people who are not overtly racist. For this reason, it is important to challenge all comments, jokes, and behaviors that promote racism. Racism is not only about immoral laws and angry mobs. Words and images matter too. I'm not suggesting that racist jokes, comments, and ideas become punishable offenses or that we police them into silence. It would probably be better in the long run to openly rob them of their power with nothing more than honesty and intelligence.

THE WORST RACISM OF ALL

Global racial disparities today are severe and dire, far worse than most people imagine, I'm sure. Right now, 1.4 billion people are trying to survive on less than US $1.25 per day.[28] Virtually all of these people are nonwhite. As previously mentioned, UNICEF reports that more than twenty-five thousand children under the age of five die every day of the year because of extreme poverty. Virtually all of these children are nonwhite. Global poverty is a complex problem and I have no reason to think that overt racism is the exclusive or even the primary cause of it. However, after many years of researching this issue, writing about it, thinking about it, and visiting some of the world's most impoverished hellholes, I have no doubt that racism *on some level* is a significant factor. I suspect that a deep unspoken racism plays a key role in this daily slaughter. I just cannot escape the nagging thought that if twenty-five thousand blonde, blue-eyed babies were dying agonizing deaths every day—that's a rate of some nine million per year—the world's wealthy white-dominated societies would find a way to end the dying sooner rather than later. Forget Klan rallies and the "N" word; today's neglect and general acceptance of dying nonwhite children is the worst racism of all. Some one hundred million nonwhite babies are scheduled to die over the next ten years, and the societies with the ability to save all or most of them show no meaningful signs of trying to do so. Yes, conferences are held, promises are made, and billions of dollars worth of aid are sent to the developing world each year. But that hasn't worked so far, and there is no reason to think the usual efforts are going to work any time soon. Meanwhile, twenty-five thousand babies died yesterday, and the same number will die again tomorrow. Nine million babies died last year. And at least the same number will die next year. If they were white, would they die?

Race belief can serve as a convenient psychological escape clause allowing North America and Europe to neglect "them," those nonwhite victims of extreme poverty. It works so well because it creates an exaggerated distance between peoples. When black- and brown-skinned babies die, it is too easy for white people to accept it, because the babies perish on the other side of a race border. Nationalistic and religious prejudices also play a major role in this, of course, but race belief allows people to ignore the suffering and death of others even within the same nation and even within the same religion. The racial health disparities in the United States, for example, are shocking and would seem to demand immediate solutions. But no, the suffering and disproportionate numbers of deaths of black, Latino, and Native American babies does not trigger a sense of alarm that compels meaningful action from the white majority. Are upper- and middle-class white people racists for allowing the dying and suffering to go on and on? Not necessarily. I would not call them racists based only on their neglect or their willingness so often to vote for leaders who clearly lack compassion. Calling them racist would conjure up images of hate-filled barbarians that may not be justified. However, these white people do fall somewhere between being full-blown racists and being fully human.

Most of them see the world through a make-believe filter of race that prevents them from being outraged over those nine million dead babies every year. The unscientific notion of race has plugged their ears so they cannot hear the cries of suffering children.

Sometimes people tell me that race consciousness is a wonderful, positive thing. Racism, for them, is a fringe problem that is confined to deviants and wholly unrelated to race belief. They scoff at the suggestion that race belief itself might be a harmful influence on the world. "There's nothing wrong with being proud of who you are," they declare. When I hear such talk, my first thought is always the price in lives that race belief indirectly claims every day. Why not be proud of being a human and seek a deeper connection with all people, including children in the developing world? When I think of racial divisions, I think of the hungry children I saw in Africa and the street children I saw in Asia, their bodies twisted and tormented by starvation and disease. If you look and listen for the members of the human family who are most in need, it is not so difficult to see through the walls of race.

RACISM

An interesting thing about racism is that the typical person on the street has ideas about this topic that are every bit as interesting—and relevant—as the experts. If I were researching quantum mechanics or continental drift, it would make sense to stick to interviewing scientists. But race belief and racism are not subjects that can be confined to laboratories and academic papers. They are mysterious and pervasive forces that entangle us all in one way or another. We should think deeply about them, and everyone should have something to say. I have asked a wide variety of people around the world to share their thoughts about race and racism. The following comments are drawn from their candid responses to my questions.

It has become unfashionable to be an overt racist in many societies today. But does this go too far in some cases? Some people may now feel that they are less able to condemn that which deserves condemnation, because such condemnation could be interpreted as racist.

"Sometimes I may feel that people who have migrated to Australia from a culture that is very different, such as the Middle East, shouldn't expect our society to bend over backwards to accommodate them because their values are so different, says Jeffrey, a twenty-four-year-old Australian. "Some people may feel it is in some way racist not to accept, or tolerate, culturally conservative views held by migrants, such as perceiving women to be second-class citizens or having creationism taught in the science classroom. I strongly feel that culturally backward positions that subvert modernity and the progression made in our developed country should not be accepted merely because many people feel it would be the politically correct thing to do, because they fear being labeled intolerant or as being racists. I feel that when people migrate to a new country, that if they have cultural views or practices that strongly contrast with that of their new country then it is up to them to at least outwardly assimilate and not expect their adopted country to radically change its principles or values to suit them."

This is a valid concern and it raises an important point. The goal of antiracists should not be to shut everyone up so that no one is ever offended. The ultimate goal should be a world in which one person can critique or even condemn the actions of another person without having the words automatically interpreted as racism. We must also be diligent about separating culture from biological race belief. These are two very different things that are often fused together in people's minds. Condemning the religion-based oppression of women, for example, is not a racist position by any stretch of the imagination.

Jeffrey believes the best way to deal with racism is to simply abandon the concept of race altogether. "However," he adds, "I don't believe this will happen for some time, because cultural evolution is a gradual process. As enough time goes by and as different groups of people from different backgrounds and cultures become integrated within society, then the superficial differences between racial groups will become increasingly minor and the traits we have in common will weaken the racial boundaries that divide us."

"I think racism still very much exists and is an extremely relevant problem in the United States and the world, despite Obama and other successful people of color," says Kimberly, a Brooklyn resident and mother of two children. "My gut reaction is that racism will always be a part of society, but with the changing demographics of the United States it is difficult to anticipate what another one hundred years will do to the dynamics of race in this country. The problem with overcoming racism is how do you do that when so much of our society's infrastructure is based on racism—it becomes a chicken and egg situation. Unless we can properly educate people of all races and provide the support needed for people to advance themselves, we will always have groups of people who are disadvantaged because of racism or at the very least from racism's past."

Like many people I speak with about race, Kimberly has a childhood memory of being hurt by racism. "I remember being treated unfairly because of my skin color," she said. "I remember not being able to play with white children on my block because their father didn't want them playing with that 'nigger child.' So of course you suffer from that as a child."

Kimberly represents an interesting mindset that is perhaps becoming increasingly common in America today. She is highly educated, sophisticated, and financially successful, as well as outwardly positive and global in her worldview. However, she is honest enough to admit that her outlook has been somewhat soured by racism in America:

I feel a kinship with people of my own race. For me, being black in America often comes with common challenges and struggles—against our history and our struggles. The history of the United States and the systematic way our country discriminates against many ethnic groups—such as with education, healthcare, and so on—influences the way I look at each racial/cultural/ethnic group. I look at race relations in the United States as White Men versus Everyone Else. So I consider myself part of Everyone Else. When I express my discontent with White Men, I do not mean to express a dislike toward individuals, but when I look at the systems under which we exist, I find myself resentful against the white establishment. I look at white men with this in mind: Arrogant, greedy and racist until proven otherwise.

Leo Igwe, thirty-eight, is a Nigerian who says race is not much of an issue for him, because he is a black man living in an all-black society. He says he once had a racist view of the world, a view not in his own favor: "When I looked at the lopsided nature of development in the world, I thought that the white race was more intelligent than the black race, but after interacting with some whites and reading about the history of other races, I began to think it has to do with nurturing not nature, that it is cultural not genetic. Given the same environment, there would be blacks that will do better than whites and vice versa."

Susan is an accomplished dancer and successful entrepreneur who connects with people on a "spiritual level that transcends race."

"I never saw or related to people based solely on race," she explains. "You grow up hearing the racial stereotypes, but I never bought into them. I always thought to myself that I had no right to define someone based solely on race if I expected others to see beyond that when they saw me. Racism has not been a problem as such for me personally, because I refuse to let it limit me. That being said, I am not naive to the fact that racism does exist. I have experienced it, but feel the only way to help change people's views on racism is to constantly challenge what they view and expect of you. I refuse to fit snuggly in the little box that they want to put me in just so that they can understand me better."

Susan admits to only one prejudice: her belief that children of interracial couples are the most beautiful children in the world. "Is that racist?" she asks.

Alex, an American Latino, says he prefers vanilla ice cream to chocolate but other than that he can't think of any "racist" tendencies he might have. He does, however, remember an ugly incident that revealed to him how much racism can hurt:

> I grew up in a small town where Latinos were half the population. I never felt or had seen racism until I was in high school playing baseball. Our team was half Latinos and half whites. We had an away game in a town called Mariposa, California. The population there was like 90 percent whites and 10 percent blacks. During the fourth inning, we began to hear people calling us names while watching the game from the parking lot near the outfield. I played center field and could hear people calling me names and cursing at me. I ignored them at first, but going back out to take the field in the sixth inning, I had enough. I turned around, and began to curse back at them. They were kids my age who hated me because of the color of my skin. Tensions got high. We were trading words back and forth, and then my coach benched me for two innings to cool off. The kids in the parking lot were asked to leave and, no surprise, we could see adults laughing in the background.
>
> It wasn't until we got back on the bus to go home my coach apologized for pulling me out because he felt I was not safe. Other team members were also upset and affected just the same, but I felt anger, hate, and [fear]. I still wanted to fight them for calling me names. I learned that racism toward me brought out a side of me I didn't like. Growing up as a Catholic, I believed that all people were made equal and I still believe

that now. I've seen some changes in the world with race. But there are still places I would never go by myself, and it takes me a while to trust anyone.

"Screw the skin color argument," says Camille. "I go with statistics. People from certain parts of the world, different countries, states, and islands *do* have certain tendencies, be they good or bad. I've come to know what to expect when dealing with certain groups of people. Some cultures are aggressive, some cultures are liberal, others are very passionate, and some are just complete asses. You can call me what you want, but you can't say I'm a liar. *However*, regardless of my views of a person's nationality or beliefs, I never treat anyone with contempt or aggression. I give everyone a chance to prove [himself or herself] as an individual to me. Sometimes I've been pleasantly surprised, and sometimes I've just had to chalk up our differences to culture and I move on. But I never hate anyone for their color—just their stupidity."

"I don't think racism will fade away," predicts Camille. "It will always be around. I just think it will be more polite. This whole race thing is a load of garbage. It's an ignorant concept. The problem with the world is the clash of cultures. There are many different cultures in the world, some weird, some backward, some aggressive, some just plain stupid. That's the problem. [There are] hundreds of different cultures and each one thinks its culture is more superior than the next. People need to feel like they are unique and special, yet at the same time they want to fit in and to belong. Over thousands of years, humans have made up cultural rituals, traditions and religions to establish themselves as a 'special breed' of people. These different beliefs and cultures have caused rifts between different groups of people throughout the world. I think the first sign of primitive organization early man came up with was segregation: 'Me color of sand....You color of wood....We no can share fire. Go find herd that look like wood. Eagle mate with Eagle. Parrot mate with Parrot.'"

Camille figures the best way to solve racism would be mass hypnotism. "You can preach acceptance [until] the cows come home, but you'll be about as successful as Nancy Reagan's 'Just Say No' to drugs campaign was. Ideally we should abandon the concept of race. But we're either thousands of years or one cataclysmic event away from that happening. We either have to forget over time or start over with a new civilization."

Kevin, an American living in New York City, says racism is here to stay. "When someone says there will one day be full equality and no such thing as color, they are dreaming. It is part of the human condition to have the need to be with your own kind. What we have to do is learn to live and work with what we have been given."

Rob is a lifetime student of advanced mathematics currently living in Florida. He can't remember ever taking the concept of race very seriously, even during his childhood years growing up in Alabama. Maybe that's because he crunched the numbers and race just didn't add up. "I refrain from generalizing without good, solid reasons that are logically demonstrable in a rigorous manner, and my mathematical training has shown me how remarkably easy it is to think that a generalization holds when it really does not," he says. "We all take mental shortcuts, whether they are perfect, imperfect, or completely outrageous. Since I perceive the idea of race as a false or poorly considered generalization, it could be said that in this respect I reject the very notion of race altogether, though I understand it exists as a rather fuzzy, highly questionable social construct."

He continues: "I believe it is important to maintain a sense that our common humanity is more fundamental than notions of race. We are all far more alike in so many more ways than we are different. Further, those differences may be celebrated as examples of the wonderful, complex variety of life."

David, a product manager for a US software company, grew up in Maine and didn't see a black person until he was around thirteen years old. "With so little exposure, race just wasn't a topic during my upbringing," he says:

> I think racism has diminished in my lifetime, and I see [fewer] and [fewer] cases of outright discrimination against other races than in the past. I think there will always be people who are unable to overcome a racist upbringing, but each generation appears to be less racist as the races continue to intermarry and socialize together. I think other divides in our society will continue to grow, such as nationalism and education levels, but I think race will not be one of the major dividing factors in the future. I think in the long term, the concept of race will naturally be abandoned as the races continue to mix. If people understood that race is more cultural than biological, this may help accelerate this process but it couldn't be mandated.

Tina, a "black" Jamaican married to a "white" man, fears she may suffer from a tiny bit of residual racism from her formative years. For example, she often struggles not to judge white people too harshly when she thinks about their fondness for engaging in dangerous activities such as diving in underwater caves and climbing very tall mountains. "Is that racist?" she asks. "I hope not, but I can't lie. I just don't understand a lot of the things white people do. I mean, come on, climbing Mount Everest? I once saw a white man on TV hanging from hooks stuck through his skin. He was trying to 'express himself,' I'm sorry, but you don't see black people doing crazy stuff like that."

"For me personally, [race] is of no consequence," declares Wendy, a "white" English journalist:

> However, the very nature of my race has conveyed very specific privileges and benefits on me—that I have not necessarily done anything to earn. I was completely oblivious to this until I was an adult and went to work as a journalist in Africa and began to realize the real meaning of the racial divides in our world. From then on, I have considered the idea of race a lot more, but not from what it means in terms of a sense of belonging but what it means in terms of oppressions and the definition of others.
>
> When I was growing up in North Yorkshire in the 1970s, the problems of unemployment were blamed on [nonwhite] immigrants by many people in the poor, white, working-class areas in which I grew up, and of course I, like many people, when I was very young, was influenced by that. However, I had parents who happened to have friends from other races and although I went to a very white school, I had a close friend whose father was Jamaican and consequently [I] wrestled to reconcile the scapegoating with my own experiences. Like many people, you don't think of your friends as the "blacks" they are talking about. By the time I was in my 20s, however, and having been to Africa, the idea that someone could be defined by the color of their skin was absurd, but I also went through the guilt when I watched *Roots* and believed that all white people were to blame for slavery.

"We're all out of Africa," Wendy continued. "One of the problems, however, is religion. If people, and many do, believe in the Bible verbatim, then God has sanctioned racism and the whole common ancestor idea, which is what should make it very easy for us all to understand [that] our commonality as a human race is completely undermined by made-up nonsense that a massive percentage of the world believes."

Seventeen-year-old Filipino student Joshua Lipan says he hopes racism fades away, but fears it will be a long time before "this irrational, bigoted collectivist view truly disappears. I think, though, that the progress that we've seen with racism is encouraging. Gone are the days when racism was taken seriously. I think it would be valuable if we accepted race as not a significant barometer for a person's value. No one race has a monopoly on talent. There are superior individuals in all races. I would divide the world by ability. Not by arbitrary physical characteristics based on race perception."

IT manager Bronwyn thinks of herself as a member of the "Australian race." While she says she feels a distance between herself and people of other races, she attributes it to cultural factors rather than biological ones:

> Some races you feel more distant [from], but that would be about core values mostly. There are some [who] have a completely different way of looking at life, of treating others or what they feel is acceptable in a given situation. Also, the races that I can't communicate with are difficult to connect with. It's a language barrier. I think growing up [in Australia] around people that coin the terms "nips," "Japs," "wogs," or others causes you to sort of go through life thinking that's normal. But then some folks will tell you it's not, so you question it. Then you live overseas and realize it's not acceptable. I don't feel any such way toward those people because it was just a name to define them. I didn't dislike them.

"I didn't grow up hearing much about black people," Bronwyn continues. "We have Australian Aboriginals, but not all of them are black, shades of brown maybe. But even if they were black, you didn't really think about their color. You thought about their actions as a race, how Aboriginals behaved, or what they did when they made music, or how they danced, or how they want money from the government. Those are the things you thought of, not their color. I don't care if they're pink: what difference does it make?"

Mike, a "white" Englishman, says, "Great strides have been achieved in the area of racial equality and racism since the 1960s, and I believe it's less of a problem today, but racism is here to stay as long as we're classified and labeled. If only there were more interracial relationships, then racism would fade away. One thing I really dislike is labels, African American, Afro-

Caribbean, American Irish, and so on. The more labels and distinctions we have, the more racial discrimination we'll have. It doesn't mean to say one has to forget where they're from; it will still be part of one's identity."

Mike says the dominant view of today's anthropologists regarding race sounds sensible and credible. "People are people! One question I've had since my first child was born is how to categorize the race of a person who is mixed. I too thought that race was a category based on color, that's what I was taught, but where does a mixed person fit in? They don't fit the norms, but society will place them in a race category anyway. My children of mixed race will be classified, in general, as black, and another mixed child, depending on appearance, will be white. So race categories are illogical. [The idea that biological races do not exist] seems so straightforward, so why all the race issues?"

George "Barefoot Man" Nowak is a popular Caribbean musician and author who spent a good part of his youth sailing around the world meeting, photographing, interviewing, and drinking with all "kinds" of people. I'm not sure, but I think he may have attained some sort of enlightened state of racism that transcends race. "Through my many travels, it's always the same," he says. "The best humans are babies and old folks. Babies are too young to hate. Old folks are usually too wise to hate or forgot who they once hated. You ask me if I'm racist? Well, if you mean racist as in prejudiced, yes I am. But I'm not prejudiced against skin color. I'm prejudiced against stupidity, ignorance, and non-commonsensical beings. When I say stupidity and ignorance, I am not talking about education—I just scratched through high school myself. Unfortunately, this planet is full of stupid people, and I don't like them. That's why I come across as a racist sometimes. I have to wonder, if there is a god, why did he put me on the wrong planet?"

A North American woman shared the following story with me. It was a trivial incident, but she wonders if it might have revealed racist thinking: "I was walking downtown and had to pass through a smaller alleyway. As I passed through, a black man came walking down the other way. For two seconds, the thought passed that this might be dangerous, and I wondered if anyone could hear me if I called out. This thought was immediately followed by the realization that this was ridiculous. If the man had been white and wearing the same clothes, would I have felt the same? I don't know."

While many people today and throughout history have used various religions as motivation and justification for their racist thoughts and behavior, Angeline, a South African, views religion as the perfect antidote to racism. "I do not consider myself to be a race," she says. "I don't make a big deal of it. I am a South African. No, race does not make me very different from other people of other races; more my belief in God, nothing to do with race. I do not consider my race to be part of my identity, even though in South Africa they do."

"There is still racism," Angeline adds. "Everywhere in the world there is racism. I have to learn to build up more tolerance for all the different types of people who I meet. It all comes back to religion again. If you put God first, no ill feelings. Eventually it becomes automatic in accepting people [for] who they are. Yes, it will always be a problem. However, I will be optimistic for the few that are prepared to look past racism."

Melanie, a 33-year-old attorney, says race is relevant to her daily life because others constantly assess her on the basis of this identification. She believes it impacts judgments about her behavior, attitude, preferences, work ethic, and family life: "I don't feel distant [from people of other races], I just have to figure out another common ground, while being careful that the other person may not be able to be comfortable with me at the same pace, or at all, because of my race, place of origin, or another issue. Race and culture are often very closely interrelated. Ancestral African traditions and practices still remain in the music, language, dress, food, and dance of predominantly black communities or countries, in adapted form."

Melanie says the only thought she has that might be considered racist is the one that tells her "a lot of white people are prejudiced against black people." She believes that race continues to be a significant problem, "I think it is here to stay. It is so ingrained in people and in society to discriminate and categorize each other. We make racist comments subconsciously without realizing it. Instinctively, I would say abandon the race concept entirely as it is too superficial and serves no constructive purpose—only to categorize and divide. [But] I simply do not see that it's possible to abolish it."

Andrea, a Canadian-Bajan-Trinidadian kindergarten teacher, says her mother shared some valuable wisdom about race with her early in life that she never forgot: "If life is a baseball game, then you're a batter with two strikes against you, even before you step up to the plate: you're black and you're a woman. Better keep your eyes on that ball."

"We were the only black family on our street [in Canada]," Andrea said. "There were only two other black kids in most of my grades at elementary school. But I never felt the sting of prejudice or racism. No one called me names or burned crosses on our front lawn."

Out with a group of students during her university days, Andrea says she asked for a window seat in the car and one of her friends raised his eyebrows and said, "Uppity negro." She says he was just joking, but the comment made her think and feel. "Those words had the power to time warp us back to the plantation days," she says:

> In that instant, I was a field hand and he was Mr. Boss Man. We were still just us, but at the same time we represented so much more—a whole history of suffering and oppression, as inescapable as the color of our skin. I went home and looked it up in the dictionary, this word, "uppity." I'd heard it before, read it in books, but now that it had been directed at me, I needed to know exactly and precisely what it meant. That definition made me smile. Yes, I thought, I am an uppity negro. No, I wasn't satisfied with my station, my place in this society; yes, I thought I deserved more. Damned right I took liberties. Isn't that what freedom means?
>
> So I embraced that word, "uppity"—made it part of me. And, since I had the dictionary in my hands, I started a search for other words that fit—words to describe what it can feel like to be black at those moments when racism rears its ugly head.

The result is the following poem, inspired by a light-hearted comment long ago. It's been bouncing around, unfinished, in her head and on paper in various versions for years. She kindly finished it so that I could include it in this book.

Uppity Negro

By Andrea Roach

Outcast, downcast and typecast
Oppressed, repressed and depressed
Refused, abused and wrongfully accused

Neglected, rejected and dejected
Slighted, blighted and uninvited
Persecuted, prosecuted and executed

Detested, divested and arrested
Framed, shamed and maimed
Segregated, denigrated and unappreciated

Disconnected, unprotected and disrespected
Brutalized, scandalized and demoralized
Unaddressed, dispossessed and laid to rest

Living... giving
Trying... vying
Coping... hoping

Satina once thought that that certain groups of people were inherently racist. "I remember being outraged as a teenager after I read Malcolm X and *The Essential Gandhi*. It really impacted the way I saw the world, and I was hurt that people would have certain judgments about me because of the way I looked."

"I don't think racism will ever truly fade away, because people like to feel different and feel superior to other groups of people," Satina says:

This need to differentiate oneself is inherent in all human beings. Plus I think many people are emotionally attached to the concept of race. I completely agree with the idea that race is created. I understood this after I read a critical review of the book *The Bell Curve*.

As I have gotten older, I understood that people are taught to discriminate. I believe racism is more nurture, not nature. It's simply socialized behavior.

Debbie, born in Guyana, says she hasn't been hurt by racism directly. "As a matter of fact," she says, "growing up in Guyana I looked Portuguese, with 'good hair,' so I was favored. People seem to be more tolerant of different races today than years ago, but racism will remain for as long as people are ignorant, continue to use it to stereotype and feed the needs of their egos."

Tauriq Moosa, a young South African writer and aspiring philosopher, must have an excellent view of the world from his Cape Town home. Just twenty-two years old, he has reasoned out far more about the reality of races than most people three and four times his age.

"I do not consider myself to be any race, as I find that as unhelpful a distinction as attributing eye color," he says. "If forced to give a description, I would put it down under the vague rubric of culture. That is, I am Indian. From both sides, I have grandparents from India, but both my parents were born in South Africa. I find it an arbitrary distinction to call oneself any race, since it just depends how far back one wishes to trace one's ancestry. Why not go all the way back to when we could first distinguish *Homo sapiens*? Why only go two or three generations back? It is particularly bizarre to me and has no impact on how I see myself or others."

Tauriq just missed growing up with the direct burdens and degradations of apartheid in South Africa. Those racist policies officially ended in 1994. His parents told him that before he was born, they couldn't buy a house they wanted near the infamous District 6 in Cape Town. Under apartheid laws, South Africa's population was strictly segregated into four groups: colored, Indian, black, and white. The area where the house was situated was designated for "colored" only, so Indians were not allowed to live there. Tauriq's parents lied and claimed to be "colored" so they could get the house. Tauriq was born the day after they moved in. He reflects on the madness of race belief:

> Not two decades ago, I would not have been able to obtain the education or work I have now—simply because my skin color was of a darker hue than my cohorts. It was an axiom for those of different shades to enter or exit by different entrances, sit on different benches, or use separate swimming pools. After F. W. de Klerk abandoned the legitimacy of apartheid —most people don't realize it was the last white president who called off apartheid, not Mandela—those axioms fell into the gutter, washed away by the stream of reason. Today you will find no "net blanke" [no black] signs, no alternating entrances. But that does not mean that there are not tendrils still clinging to the edge of gutters. Our "rainbow nation," as the great Desmond Tutu calls it, is one whose iridescence is fading to black. Not the black race but one of a line drawn across a page, demarcating where we cross into troubled territory.

Tauriq favors abandoning the concept of biological race. "It must be abandoned," he said:

> I have never in my life dealt with it—as in I have never [based decisions on race]. As people are more important than their race and their religion—two concepts, consequently, which I believe should be buried in the graveyard of bad ideas—we can further peel the veneer of past impositions to reveal the living ghost of humanity that lurks behind. We are afraid of these ghosts, it seems. We are inherently fearful of other people and do anything we can to demarcate where I stand, where they stand. Like chalk outlines, we have circles where the dead ideas lay and we still treat them as though they play a factor in our lives. But they are just chalk outlines of where bad ideas used to dwell. It is time to rub out those circles and move on.

Carolina, a thirty-three-year-old Brasilian (she insists that I spell it with an "s"), has a lot to say about race. She is thoughtful and, as a Red Cross employee and habitual volunteer for good causes, she instinctively tackles anything that may have negative consequences for innocent people. Yes, this woman was destined for a head-on collision with the concept of race. Her comments, I feel, will be enlightening for anyone who has grown up in a society where race seems natural and logical:

> In the United States, if you have one drop of black blood you are considered black or mixed. In Brasil, things aren't like that, and I struggled trying to find my identity [while living in the United States] until the point that I realized that it was something I was doing for others and not really for myself. I never met my grandparents on my father's side, but on my mother's side my grandfather was black, and my grandmother is as mixed as can be: Native Brasilian, European, and African. My mother identifies herself as being black, and while I hear that my maternal grandmother was in fact black, my father identifies himself as being white.
>
> Living in the United States and having to take a crash course on becoming a minority race did become an important part of my identity. It wasn't something that I made an issue to distinguish for myself, but it was the one thing that made me different in a school where the majority of the kids were white. Yet for me, becoming "Latina," something which I wasn't until after I left Latin America, was not so much about race as it was about culture and perspective. But even that came with hiccups: in

college I was told that I was not Latina because my first language was Portuguese and not Spanish, and thus I did not "qualify." In the States, there is such pressure to define yourself and wear a label. People like me, who aren't so definable or willing to be confined, end up creating a bit of a chaos, which at first leads to discussions but later leads only to frustration and exasperation, as others can't wrap their heads around why it is that I can't just "pick a side." But how can you pick a side when your cousins have nicknames that range from "Neguinho"—little black one— to "Alemaozinho"—little German one?

"I would like to think that as we become more intertwined and more mixed that we will do the sensible thing and drop this need to categorize and define and separate," says Carolina:

It makes complete sense to me that race is a figment of our collective imagination. I never understood how pigmentation or the shape of one's features made one more or less able to perform academically, for example. It is funny, though, that of all the things which we, the human race, have outgrown in our development and evolution—the move towards democratic societies based on capacity and merit and not birth inheritance, recognition that education should be for men and women, banning of cruel and inhumane torturous methods, such as [tarring] and feathering, for example —race is something that has stuck and has yet to be debunked. One would think that by now, given all that we know, more questions would've risen to challenge this notion. Yet we don't [challenge it]. Why? Because to destroy such a notion would shake the very foundation that the status quo rests on. It would mean judging people truly based on their own merits and not on the facade that we created to ensure they "know their place" in this world. The fiction of race has imposed limitations on human beings for centuries. As Steven Biko said, "The most potent weapon in the hands of the oppressor is the mind of the oppressed."

Sarah, a twenty-seven-year-old Canadian, doesn't have much use for race. "I don't believe my race is an important part of my identity. There doesn't seem to be a unifying force behind the term 'Caucasian.' I have no innate sense of pride in 'what we as a race have accomplished!' If anything, there is guilt attached to the many crimes against humanity that Caucasians have committed in the past thousand years. Slavery, oppression, theft, murder on an international scale. It doesn't matter that I'm Canadian, with no direct linkage to these injuries. I still feel shame when I see documentaries, read historical accounts, and listen to modern day news."

Sarah was an "active campaigner against racism" within her family: "My grandfather immigrated from Britain in the early '60s because he felt that his small community was being taken over by Africans. He remained very vocal and strongly against people of a wide variety of different skin colors and creeds. I lived with my grandparents as a teenager, and my temper would always flare terribly whenever he watched the news and hurled racial epithets at people of color whose pictures were shown during the nightly news. I would rage against his racism and even once told him that 'people like him were dying out' and our new generation would not be so blind and ignorant as his. He just laughed at me. I think that sometimes he just liked seeing my blood boil."

"It's sad to say, but racism has not really become less of a problem in my lifetime," says John, a "white" Englishman living in Manchester:

> Progress has been made in some areas of life, but people will always be bigoted and discriminate in some ways, because humans often have ugly views about things. They are just made that way. People are always looking for others to blame for things, and it's easy to target someone who looks or acts differently.
>
> Of course there are physical differences in peoples around the world, so biological diversity is a reality. But it is different cultures and creeds that often lie behind the real clashes, troubles, and misunderstandings in the world. It has always been that way and probably always will [be].

"I have experienced being on the receiving end of the ugliness of direct racism when for a time I lived in a predominantly Jamaican area of a town in the West Indies," says John. "Most people there, like myself, had a relaxed live-and-let-live attitude to life. There were others, however, who appeared to go out of their way to make me feel uncomfortable and unwelcome, and that made me angry. But there will always be people who, to make up for their own shortcomings, look to blame someone else for their situation."

"It would be best for everyone to learn to accept all races and to tolerate racial differences," John continues. "But frankly, that is unlikely to happen. Humans, with a few outstanding exceptions, deep down—and sometimes not so deep down—are often just nasty, greedy, selfish, and downright unpleasant creatures toward each other and toward other, more worthy animals that share this planet."

Stacey and her husband wanted to have a child. After years of trying, however, it became clear to them that it just wasn't going to happen. This "white" American couple discussed adoption and, after a lot of research, decided to go that route. It wasn't easy. They took classes geared to becoming foster parents and eventually adopting but were told that there would be no guarantees, even after a child had lived with them for years.

"It's such a shame," says Stacey, "that there are so many children out there that need homes and they [the government and the adoption agencies] make it so difficult to adopt."

Frustrated with this situation, they found an adoption agency that seemed promising. After filling out tons of paperwork and having every aspect of their lives investigated, they were cleared to adopt. This is the point where race enters the story: "They had a 'special needs' adoption program [for children with physical and/or mental handicaps] and we came to find out that any child that is African American is also considered 'special needs.' We couldn't believe it! The child can be perfectly healthy but is considered special needs just because he is black [and therefore difficult to place with a family]. They explained that it's because there are so many black kids in the system and most of the people who adopt want perfect little white babies."

"It was crazy to us," says Stacey. "We told them that we didn't care what race the child was. We wanted to be parents and we loved all children regardless of the color of their skin."

Eventually it all came together for Stacey, her husband, and one little black baby who needed a home and a family.

"He is the most beautiful thing that has ever happened to me," she says. "Today, people say all the time how lucky he is to have us. But I always correct them and say that *we* are the lucky ones. He is just finishing up his first-ever T-ball season. It goes by so quickly. Soon he'll be a first grader! He is such a smart kid, and he loves to laugh. I'm so very proud of him."

Stacey and her husband have no illusions about the challenges ahead as their son grows up. There will be lots of questions. And while they say they do not duck the subject of race or his biological mother, they do want him to see people first and recognize that love matters more than race.

"It amazes me each day that he is so unaffected by our [racial] differences," says Stacey. "He'll share his [adoption] story with people occasionally, and he never mentions the difference in our races, which is exactly what I want for him. We see people as people and we embrace our differences."

Kellie, twenty, is a college student in Swaziland, Africa. She has traveled extensively, visiting more than thirty countries, and says she is who she is today because of the variety of people she has befriended along the way. She doesn't like the term "race":

> The idea of labeling yourself as one particular race is like putting yourself into a very simplistic container that has an strong affinity for conflict. People like to stick to what's safe, and more often than not this means that people of the same race form friendships with each other as they are more similar and share common ideals, and religious practices etcetera. However, these barriers create room for damage. If people don't remove themselves from the safe area and venture out to find similarities within people of other races, the boundaries will always exist.
>
> I have become much more aware of racism as I've grown older and the problems associated with it. I find it disheartening to see the emphasis people place on the important of race, and how it affects relationships.

This awareness has not left Kellie without hope for the future, however. "I feel racism is becoming slightly less of a problem," she says, "solely due to the modern ideals people are adopting and the fact that people of different races are being forced to depend on each other more than ever before."

FACING DOWN THE BEAST

It is important to understand that people who are strongly racist are not necessarily mentally ill or in any way defective human beings. Racism is not an illness; it's a destructive worldview based on faulty reasoning. Some of the most intelligent people who have ever lived have been flaming racists. It may be a stupid way to think, but racism doesn't mean one is necessarily a stupid person. Thomas Jefferson, for example, was undeniably a brilliant man—and yet he was terribly racist. Yes, when I was wading through the madness of that Ku Klux Klan rally, it would have been easy to just shake my head and declare everyone present insane. But that wouldn't have been true. In most cases, overt racists are doing nothing more than acting on their racial beliefs, taking them to the extreme. The belief is the source of the problem. And that belief is what should be addressed if we are ever to have any hope of outgrowing racism. Getting angry at racists and thinking of them as mutants is not productive. Those who say we should all love or at least accept one another and oppose racism because it hurts people are only fighting half of the battle. Yes, there is a moral component to racism, but that is not all there is to it. We have to try to understand as much as we can about racism and defeat it, not just with our hearts but with our intellects as well. We are all vulnerable to stumbling and making mistakes in the way we view the world and our fellow humans. Thinking in stereotypes and tripping over irrational beliefs is a common human problem. "We shouldn't treat prejudice as pathological just because it offends us," says anthropologist Francisco Gil-White, "If we aim to transcend ethnic strife, we would be wise to understand the role that perfectly normal human psychology plays in producing it."[29]

"The challenge facing us is to confront our true nature," declares a bold editorial in *New Scientist* magazine: "Instead of denying that our tendency to prejudice exists, we would do well to understand why and when it is most likely to be triggered. This might give us the chance to set aside the urge to crudely pigeonhole people, and instead deal with them as individuals. Such behavior is certainly more constructive and civilized, and it stands to improve our success as social, political and business animals. It would be naive to suppose that such self-knowledge will instantly dissolve the deep-seated prejudice that exists around the world, but it is a start."[30]

It is not unusual at all for life-forms to gravitate toward life-forms that are closely related to them and may give them safer, preferential treatment. Even plants do it. At least four species of plants have the ability to sense the presence of a nearby plant and determine if it is a relative or not. If it's not related, the plant will sprout new roots to absorb more nutrients from the soil. If it is kin, however, the plant will restrain itself, share the soil, and allow the other plant a better chance of success.[31] But we aren't plants, and *all* people are our kin. Furthermore, we can think through and beyond whatever prejudices and instinct toward favoritism nature may have burdened us with. Rightly or wrongly, most people probably do tend to trust those who look similar, because they are perceived to be more closely related. This might have made some sense thousands of years ago when we were roaming dangerous landscapes without cell phones. Unfortunately, this urge has been pushed to extremes with race belief in recent centuries. We also assume that different languages, dress, and other cultural traits are inseparable from imagined biological race categories, thereby magnifying the significance of biological diversity. If we think about it, it's not difficult to recognize that it makes little sense to trust and feel closer to a stranger based on race alone. But this is what race belief asks us to do. White people rob, harm, and kill white people every day around the world. Black people rob, harm, and kill black people every day around the world. Hispanics and Asians harm other Hispanics and Asians as well. Regardless of what you may have been taught in childhood, there is no guarantee of safety to be found within race. Your security comes down to being around good people and avoiding bad people.

For a long time, scientists believed that our brains automatically identified individuals by age, sex, and also race. This was an instinctive reaction that had evolved many thousands of years ago, or so the thinking went. We see race because our minds are hardwired to do so. However, a fascinating study headed by University of Pennsylvania psychologist Robert Kurzban has cast doubt on the idea that the human mind reads race automatically.[32] Kurzban showed test subjects a series of images of basketball players representing two teams. A statement was presented as each image was shown. The images were all sequential and were set up as if there had been an altercation during a game. After a brief distraction exercise, the test subjects were asked to match the players with their statements. It was tough to do, and many mistakes were made. But guess what? The mistakes told a lot. Test subjects were better at matching a statement to the jersey color of the player who made the statement than they were at matching a statement to the race of the player. They remembered shirt colors—team affiliations—better than they remembered race. The experiment suggests that artificial teams or coalitions can easily supersede "natural" races in the "natural" mental assessments of the test subjects. This is not a good result for committed race believers, who see racial recognition as some deeply entrenched component of the mind.

In a summary of the experiment, Dr. Kurzban states that "no part of the human cognitive architecture is designed specifically to encode race. We hypothesize that the (apparently) automatic and mandatory encoding of race is instead a byproduct of adaptations that evolved for an alternative function that was a regular part of the lives of our foraging ancestors: detecting coalitions and alliances.... contrary to prior claims, race is not inevitably encoded with equal strength across social contexts. In a social world where the active coalitions are easy to encode and do not track race—even briefly—encoding by race decreases."[33]

Dr. Kurzban concludes: "What is most striking about these results is just how easy it was to diminish the importance of race by manipulating coalition.... The sensitivity of race to coalitional manipulation lends credence to the hypothesis that, to the human mind, race is simply one historically contingent subtype of coalition. Our subjects had experienced a lifetime in which ethnicity (including race) was an ecologically valid predictor of people's social alliances and coalitional affiliations. Yet less than four minutes of exposure to an alternative social world in which race was irrelevant to the prevailing system of alliance caused a dramatic decrease in the extent to which they categorized others by race. This implies that coalition, and hence race, is a volatile, dynamically updated cognitive variable, easily overwritten by new circumstances. If the same processes govern categorization outside the laboratory, then the prospects for reducing or even eliminating the widespread tendency to categorize persons by race may be very good indeed."[34]

Let's make sure this is clear: test subjects had an easier time remembering team affiliation than racial affiliation. They forgot race with ease, although we have been led to believe that race is innate and of immense importance to the human mind. In general, people seem to pick up and remember gender and age with much greater consistency than race. Racial identification, it seems, may not be ingrained or innate at all. I asked Dr. Kurzban to comment further on the popular notion that our minds are evolved or hardwired to naturally identify the races of individuals. This seems to me to be a fundamental issue that has profound implications about how societies might try to eradicate racism.

"I think 'hardwired' is nearly always a bad description of the way minds work," Dr. Kurzban said: "A better way to conceptualize this issue is to ask if there are mechanisms of the mind designed to categorize by race. I think the answer to that is no, and my work, and follow-up work by others—I think—supports this view. In terms of getting rid of [racism], I think policy makers have this job to do. As a scientist, what I can say is that people categorize people using many different kinds of cues, and the categorization system looks quite flexible. In our studies, we highlight dimensions orthogonal [unrelated] to race, and that is what seems to produce the effects we observe."[35]

Dr. Kurzban added that further research is needed, because race is "an important issue and scientific progress can help us understand how to achieve policy goals."[36]

Finally, he made the point that he does not see attention to race as a necessary component of a society or an individual's thought processes. "People can categorize others in many different ways, and in that sense, I don't think that there is an inevitability to any of these ways," he said.[37]

University of Michigan psychologist Lawrence Hirschfeld believes that "our minds seem to be organized in a way that makes breaking the human world into distinct groups almost automatic."[38] But a natural urge to divide and categorize ourselves does not mean that we are all born to be racist or even race conscious. "Distinct groups" does not have to mean races. Nor do our groups necessarily have to be in conflict. If we must have groups, whatever they may be, there is no reason we can't insist that they must be more sensibly constructed than races and that they cooperate toward some greater good for all. For example, the serious environmental problems facing the world right now defy compartmentalization or any notion of limited effect. Everyone has a stake in biodiversity loss, dying oceans, climate change, drinking water shortages, and so on. It doesn't matter if we can't resist categorizing ourselves one way or another. What matters is that we avoid inflating the importance of our groups with false beliefs about the abilities and limitations of members, so that we can come together, maximize our abilities, and find solutions.

SOMEBODY HAS TO DO IT

One of the pleasures of writing this book was that it led me to Jane Elliott. Back in the 1960s, she was a third-grade schoolteacher in Riceville, a small all-white Iowa town. When Martin Luther King Jr. was shot in April 1968, her students were confused. "Why did they shoot that King?" a boy asked her. The class had just recently made King their "Hero of the Month." Elliott decided to do something meaningful, to teach her class about prejudice, but in a way that they might remember. After discussing the murder of King and racism in general, Elliott asked the class if they would be willing to take part in a special lesson. They said yes.

Elliott wanted the children to learn about prejudice by experiencing it. She split the class in two based on eye color. On the first day, those with blue eyes would be the "superior" students. Brown-eyed students would be "inferior" and have to wear a paper collar. Elliott favored the blue-eyed students throughout the day, giving them a longer recess time and allowing them to be first in line for lunch. She told the brown-eyed students that their classwork was not good enough and praised the blue-eyed students for their efforts. The next day she reversed the roles.

Elliott was shocked by the impact the two-day exercise had on her class. In a short time, the "inferior" students began acting and looking as if they really were inferior, and many of the "superior" students became mean and enthusiastically embraced their dominant status. "I watched what had been marvelous, cooperative, wonderful, thoughtful children turn into nasty, vicious, discriminating little third-graders in a space of fifteen minutes," Elliott said. She recognized that she had "created a microcosm of society in a third-grade classroom." When it was over, the students were different. "The kids said over and over, 'We're kind of like a family now.' They found out how to hurt one another and they found out how it feels to be hurt in that way and they refuse to hurt one another in that way again."[39]

Elliott says that after she had performed the exercise in her class for four years, a University of Northern Iowa professor conducted a study and found that students who had been in her class were less racist in their attitudes than other students in the school. Even more interesting, however, was the discovery that *all* of the students at the school tended to be less racist in their attitudes than other students in the community.[40]

Elliott's students not only retained the memory of that unique classroom exercise but it seems that they also somehow influenced their peers in other classes to be less racist as well. It's nice to know that kindness and cooperation are contagious too.

Now in her seventies, Elliott is still going strong, teaching her lesson to many thousands of young people and adults around the world in workshops for schools, government agencies, and private companies. But it never gets any easier. "Every time I do it I end up with a migraine headache. I absolutely hate this exercise," she told a PBS interviewer. "But the worst of it is that the exercise is as necessary today as it was in 1968."[41]

I tracked down Jane Elliott, and I am happy to report that she is no less fired up against racism today than she was in 1968. "There is still a need for what I do, because we are still conditioning people all over the world to the myth of different races and to the worse myth of the superiority of whiteness," she said: "The reason we haven't made greater progress in understanding the myths is that those who are in positions of power are, in most cases, white males who actually believe that their skin color and their gender do, indeed, make them superior to the rest of us and, therefore, most deserving of possessing and wielding power. I keep on doing what I'm doing because the racists keep on doing what they're doing. When they stop, so will I. But I'll still have a job where sexism and ageism and homophobia and ethnocentrism are concerned, won't I?"[42]

Some critics say Elliott is too confrontational, too rude. No, she's not too rude. She's not as rude as a lynching or allowing nine million nonwhite babies to die in poverty every year. I applaud her antiracist anger and her willingness to push back. But I'm also frustrated. Do we really have to experience degradation and inequality in a classroom exercise or company workshop in order to recognize that that these are bad things and that they exist all around us? Do we really need to have someone separate us by eye color or some other trivial trait and berate us for a day to get us to care about discrimination? I hope not. I hope we can start doing a better job of figuring this stuff out on our own.

The writing is on the wall. Ashley Montagu put it up there seventy years ago: Biological races do not exist beyond the confines of our minds. This fantasy has caused enough damage; it's time to let it go. The crucial first step is to see races for what they really are, imaginary boxes filled with members of the human family who should feel much closer to us but cannot because of race belief. We have to be like that kid in the fairy tale who spoke up and said, "Hey, the emperor is not wearing any clothes!" If we are honest enough and courageous enough to confront not only racism but race belief itself, we just might find that we can create a better world. Not a world cheapened by less diversity, but a world made richer by tearing down the false walls we built.

NOTES

1. Steven Stavropoulos, *The Beginning of Wisdom* (New York: Marlow, 2003), p. 52.

2. Mark Twain, "Concerning the Jews," *Harper's Magazine* (March 1898).

3. Nelson Mandella, *Long Walk to Freedom* (Boston: Little, Brown, 1994), p. 542.

4. Marissa, interview by author, April 25, 2008.

5. Jared, interview by author, April 27, 2008.

6. "Gang War: Bangin' in Little Rock," directed by Mark Levin, HBO, 1994.

7. Associated Press, "Ahmadinejad Dropped Holocaust Denial from Speech," *San Francisco Chronicle*, April 21, 2009, www.sfgate.com/cgi-bin/article.cgi?f=/n/a/2009/04/19/international/i084035D56.DTL (accessed September 15, 2009).

8. Guy P. Harrison, "I Was Preparing Myself to Die," *Caymanian Compass*, December 19, 2003, pp. A24–A25.

9. Guy P. Harrison, "Stronger Than Evil," *Caymanian Compass*, March 4, 2004, pp. A25–A26.

10. Guy P. Harrison, "Defying Hitler's Evil," *Caymanian Compass*, August 8, 2003, pp. A16–A17.

11. Guy P. Harrison, "Embraced by Evil," *Caymanian Compass*, December 5, 2002, pp. 15–16.

12. Guy P. Harrison, "From the Nazis to NASA," *Caymanian Compass*, September 27, 2002, p. A19.

13. See Stanley Milgram, *Obedience to Authority: An Experimental View* (New York: Perennial Classics, 2004); and Philip Zimbardo, *The Lucifer Effect: Understanding How Good People Turn Evil* (New York: Random House, 2008).

14. Iris Chang, *The Rape of Nanking* (New York: Penguin Books, 1997).

15. Guy P. Harrison, "Defy the Divine Wind," *Caymanian Compass*, May 3, 2002, pp. A22–A23.

16. Guy P. Harrison, "The Man Who Made Peace," *Caymanian Compass*, June 20, 2003, pp. A18–A20.

17. CNN Transcripts, air date February 8, 2009, 22:00, http://transcripts.cnn.com/TRANSCRIPTS/0902/08/cnr.05.html (accessed September 15, 2009).

18. Nick Wynne, interview by the author, April 6, 2009.

19. Ibid.

20. Dan DeWitt, "Past Pain Still Present," *St. Petersburg Times*, July 5, 2005, www.sptimes.com/2005/07/05/Hernando/Past_pain_still_prese.shtml (accessed September 15, 2009).

21. Southern Poverty Law Center, "Hate Group Numbers Up by 54% Since 2000," February 26, 2009, http://www.splcenter.org/news/item.jsp?aid=366 (accessed September 15, 2009).

22. Eldridge Cleaver, *Soul on Ice* (New York: Delta, 1999), p. 84.

23. Implicit Association Test (IAT), https://implicit.harvard.edu/implicit/ (accessed September 15, 2009).

24. Ibid.

25. University of Illinois at Urbana–Champaign, "Negative Perception of Blacks Rises with More News Watching, Studies Say," *ScienceDaily*, July 17, 2008, http://www.sciencedaily.com/releases/2008/07/080717134527.htm (accessed September 15, 2009).

26. Ibid.

27. Brent Staples, "The Ape in American Bigotry, from Thomas Jefferson to 2009," *New York Times*, February 27, 2009, http://www.nytimes.com/2009/02/28/opinion/28sat4.html (accessed September 15, 2009).

28. Peter Singer, *The Life You Can Save* (New York: Random House, 2009), p. 7.

29. Quoted in Mark Buchanan, "Are We Born Prejudiced?" *New Scientist*, March 17, 2007, www.newscientist.com/article/mg19325952.000-are-we-born-prejudiced.html?full=true (accessed September 15, 2009).

30. Editorial, "We Must Face Our Prejudicial Urges," *New Scientist*, March 17, 2007, p. 5.

31. Carol Kaesuk Yoon, "Loyal to Its Roots," *New York Times*, June 10, 2008, www.nytimes.com/2008/06/10/science/10plant.html (accessed September 15, 2009).

32. Robert Kurzban, John Tooby, and Leda Cosmides, "Can Race Be Erased? Coalitional Computation and Social Categorization," *Proceedings of the National Academy of Sciences* 98, no. 26 (December 18, 2001), www.pnas.org/content/98/26/15387.full (accessed September 15, 2009).

33. Ibid.

34. Ibid.

35. John Kurzban, interview by the author, April 20, 2009.

36. Ibid.

37. Ibid.

38. Buchanan, "Are We Born Prejudiced?"

39. PBS, "A Class Divided," *Frontline*, 1985, www.pbs.org/wgbh/pages/frontline/shows/divided/etc/synopsis.html (accessed September 15, 2009).

40. Jane Elliott, interview by PBS, "An Unfinished Crusade," *Frontline*, December 19, 2002, http://www.pbs.org/wgbh/pages/frontline/shows/divided/etc/crusade.html (accessed September 15, 2009).

41. Ibid.

42. Jane Elliott, interview by the author, April 25, 2009.

SELECT BIBLIOGRAPHY

Alland, Alexander. *Race in Mind: Race, IQ and Other Racisms.* New York: Palgrave, 2002.

Allport, Gordon W. *The Nature of Prejudice.* Cambridge, MA: Perseus Books, 1979.

Aronson, Marc. *Race: A History beyond Black and White.* New York: Ginee Seo Books, 2007.

Asim, Jabari. *The N Word.* Boston: Houghton Mifflin, 2007.

Back, Les, and John Solomos. *Theories of Race and Racism.* New York: Routledge, 2000.

Baker, Lee D. *From Savage to Negro: Anthropology and the Construction of Race, 1896–1954.* Berkeley: University of California Press, 1998.

Bartholomew, David. *Measuring Intelligence: Facts and Fallacies.* Cambridge: Cambridge University Press, 2004.

Brace, C. Loring. *"Race" Is a Four-Letter Word: The Genesis of a Concept.* New York: Oxford University Press, 2005.

Cavalli-Sforz, Luigi Luca, and Francesco Cavalli-Sforz. *The Great Human Diasporas: The History of Diversity and Evolution.* New York: Helix Books, 1995.

Chang, Iris. *The Rape of Nanking.* New York: Penguin Books, 1997.

Cleaver, Eldridge. *Soul on Ice.* New York: Delta, 1999.

Dalmage, Heather. *Tripping on the Color Line: Black-White Families in a Racially Divided World.* New Brunswick, NJ: Rutgers University Press, 2000.

Dalton, C. H. *A Practical Guide to Racism.* New York: Gotham Books, 2008.

Diamond, Jared. *Guns, Germs, and Steel.* New York: Norton, 1997.

———. *The Third Chimpanzee: The Evolution and Future of the Human Animal.* New York: Harper, 1992.

Ehrlich, Paul. *Human Natures: Genes, Cultures, and the Human Prospect.* New York: Penguin Books, 2002.

Fish, Jefferson M., ed. *Race and Intelligence: Separating Science from Myth.* Mahwah, NJ: Erlbaum, 2002.

Flynn, James. *What Is Intelligence? Beyond the Flynn Effect.* Cambridge: Cambridge University Press, 2009.

———. *Where Have All the Liberals Gone? Race, Class, and Ideals in America.* New York: Cambridge University Press, 2008.

Fraser, Stephen, ed. *The Bell Curve Wars.* New York: Basic Books, 1995.

Gallagher, Charles. *Rethinking the Color Line.* New York: McGraw Hill, 2007.

Gladwell, Malcolm. *Outliers: The Story of Success.* Boston: Little, Brown, 2008.

Gossett, Thomas F. *Race: The History of an Idea in America.* New York: Oxford University Press, 1997.

Gould, Stephen Jay. *Ever since Darwin.* New York: Norton, 1997.

———. *I Have Landed.* New York: Harmony Books, 2002.

———. *The Richness of Life: The Essential Stephen Jay Gould.* New York: Norton, 2006.

Graves, Joseph. *The Race Myth: Why We Pretend Race Exists in America.* New York: Dutton, 2004.

Graves, Joseph L. *The Emperor's New Clothes: Biological Theories of Race at the Millennium.* New Brunswick, NJ: Rutgers University Press, 2001.

Greene, Eric. *Planet of the Apes as American Myth: Race, Politics, and Popular Culture.* Jefferson, NC: McFarland, 1996.

Griffin, John Howard. *Black like Me.* New York: New American Library, 2003.

Hoberman, John. *Darwin's Athletes.* Boston: Mariner, 1997.

Johanson, Donald, and Kate Wong. *Lucy's Legacy: The Quest for Human Origins.* New York: Harmony Books, 2009.

Jordan, Michael. *Driven from Within.* New York: Atria, 2005.

Jordan, Winthrop D. *The White Man's Burden: Historical Origins of Racism in the United States.* New York: Oxford University Press, 1974.

Kida, Thomas. *Don't Believe Everything You Think: The Six Basic Mistakes We Make in Thinking.* Amherst, NY: Prometheus Books, 2006.

King, Coretta Scott. *The Words of Martin Luther King, Jr.* New York: Newmarket, 1983.

Kohn, Marek. *The Race Gallery.* London: Vintage, 1996.

Lewontin, R. C., Steven Rose, and Leon Kamin. *Not in Our Genes.* New York: Pantheon Books, 1984.

Lewontin, Richard. *Human Diversity*. New York: Scientific American Library, 1982.

Lynch, John, and Louise Barrett. *Walking with Cavemen*. New York: DK, 2003.

Malik, Kenan. *Strange Fruit: Why Both Sides Are Wrong in the Race Debate*. Oxford: Oneworld, 2008.

Marger, Martin. *Race and Ethnic Relations: American and Global Perspectives*. Belmont, CA: Wadsworth, 2009.

Marks, Jonathan. *Human Biodiversity: Genes, Race, and History*. New York: Aldine de Gruyter, 1995.

———. *What It Means to Be 98% Chimpanzee*. Berkeley: University of California Press, 2002.

Miele, Frank. *Intelligence, Race, and Genetics: Conversations with Arthur Jensen*. Boulder, CO: Westview, 2002.

Milgram, Stanley. *Obedience to Authority: An Experimental View*. New York: Perennial Classics, 2004.

Molnar, Stephen. *Human Variation: Races, Types, and Ethnic Groups*. Upper Saddle River, NJ: Prentice Hall, 1998.

Montagu, Ashley. *Man's Most Dangerous Myth: The Fallacy of Race*. Walnut Creek, CA: AltaMira, 1997.

———, ed. *Race and IQ*. New York: Oxford University Press, 1999.

Murdoch, Stephen. *IQ: A Smart History of a Failed Idea*. Hoboken, NJ: John Wiley and Sons, 2007.

Myrdal, Gunnar. *An American Dilemma: The Negro Problem and Modern Democracy*. New York: Harper, 1944.

Nisbett, Richard E. *Intelligence and How to Get It*. New York: Norton, 2009.

Olson, Steve. *Mapping Human History*. Boston: Houghton Mifflin, 2002.

Omi, Michael, and Howard Winant. *Racial Formation in the United States*. New York: Routledge, 1994.

Rattansi, Ali. *Racism: A Very Short Introduction*. Oxford: Oxford University Press, 2007.

Rediker, Marcus. *The Slave Ship: A Human History*. New York: Viking, 2007.

Reed, Annette Gordon, ed. *Race on Trial*. Oxford: Oxford University Press, 2002.

Root, Maria P. P. *Love's Revolution: Interracial Marriage*. Philadelphia: Temple University Press, 2001.

Sachs, Jeffrey D. *The End of Poverty: Economic Possibilities for Our Time*. New York: Penguin, 2005.

Sagan, Carl. *Billions and Billions: Thoughts on Life and Death at the Brink of the Millennium*. New York: Random House, 1997.

Sargent, Lyman Tower. *Extremism in America*. New York: New York University Press, 1995.

Sarich, Vincent, and Frank Miele. *Race: The Reality of Human Differences*. Boulder, CO: Westview, 2004.

Sawyer, G .J., and Viktor Deak. *The Last Human: A Guide to Twenty-Two Species of Extinct Humans.* New Haven, CT: Yale University Press, 2007.

Schaffer, Richard T. *Racial and Ethnic Groups.* 11th ed. Upper Saddle River, NJ: Pearson Education, 2007.

Selig, Ruth, Marilyn London, and Ann Kaupp. *Anthropology Explored.* Washington, DC: Smithsonian Books, 2004.

Seuss, Dr. *The Sneetches and Other Stories.* New York: Random House, 1961.

Shermer, Michael. *Why Darwin Matters.* New York: Times Books, 2006.

Shipman, Pat. *The Evolution of Racism.* New York: Simon & Schuster, 1994.

Singer, Peter. *The Life You Can Save: Acting Now to End World Poverty.* New York: Random House, 2009.

Smith, Cameron M., and Charles Sullivan. *The Top 10 Myths about Evolution.* Amherst, NY: Prometheus Books, 2007.

Tattersall, Ian. *The Human Odyssey.* New York: Prentice Hall, 1993.

Tattersall, Ian, and Jeffrey H. Schwartz. *Extinct Humans.* Boulder, CO: Westview, 2001.

Terkel, Studs. *Race.* New York: New Press, 1992.

Thomas, Hugh. *The Slave Trade.* New York: Simon & Schuster, 1997.

Twine, France Winddance. *Racism in a Racial Democracy: The Maintenance of White Supremacy in Brazil.* New Brunswick, NJ: Rutgers University Press, 1998.

Wade, Nicholas. *Before the Dawn: Recovering the Lost History of Our Ancestors.* New York: Penguin Books, 2006.

Walvin, James. *A Short History of Slavery.* New York: Penguin, 2007.

Webster, Yehudi. *The Racialization of America.* New York: St. Martin's Press, 1992.

Wells, Spencer. *The Journey of Man: A Genetic Odyssey.* Princeton, NJ: Princeton University Press, 2002.

Wilson, Willima Julius. *More Than Just Race: Being Black and Poor in the Inner City.* New York: Norton, 2009.

Wise, Tim. *White like Me: Reflections on Race from a Privileged Son.* New York: Soft Skull, 2009.

Woodward, C. Vann. *The Strange Career of Jim Crow.* New York: Oxford University Press, 2002.

Wright, Kai, ed. *The African American Experience: Black History and Culture.* New York: Black Dog and Leventhal Publishers, 2009.

Zimbardo, Philip. *The Lucifer Effect: Understanding How Good People Turn Evil.* New York: Random House, 2008.

RECOMMENDED WEB SITES

The American Anthropological Society's Statement on Race:
   www.aaanet.org/stmts/racepp.htm

American Association of Physical Anthropologists' Statement on Biological Aspects of Race:
   http://physanth.org/association/position-statements/biological-aspects-of-race/

The American Anthropological Society's Statement on Race and Intelligence:
   www.aaanet.org/stmts/race.htm

The American Anthropological Society's Statement on the Misuse of Scientific Findings to Promote Bigotry and Racial and Ethnic Hatred and Discrimination:
   www.aaanet.org/stmts/bigotry.htm

Understanding Race (a project of the American Anthropological Society):
   www.understandingrace.org

Race: The Power of an Illusion (companion site to documentary):
www.pbs.org/race

Becoming Human (human evolution):
www.becominghuman.org

Is Race "Real"?:
http://raceandgenomics.ssrc.org/

Unnatural Causes...Is Race Making Us Sick? (companion site to documentary)
www.unnaturalcauses.org

RaceSci (history of race and science):
www.racesci.org

Genetic Anthropology, Ancestry, and Ancient Human Migration (by the Human Genome Project):
www.ornl.gov/sci/techresources/Human_Genome/elsi/human migration.shtml

Southern Poverty Law Center (tracks racist groups in the United States):
www.splcenter.org/index.jsp

INDEX